Table of Contents

"Portion of this work first appeared in the Indian Journal of Advanced Physics, published under the title *'Introduction to Universal Law to Provide Reasons for the Birth and Working of the Universe', Indian Journal of Advanced Physics*, 5(1), 11–16. https://doi.org/10.54105/ijap.a1059.05010425

Made in India.

This work is a presentation of theoretical research and is not intended as a definitive scientific textbook until validations are received.

Preface

There are moments in every person's life when a thought strikes so deeply, it becomes impossible to ignore. For me, that thought was not fleeting; it became a question, a burden, a calling. The laws of physics, the universe we inhabit, the nature of time, death, and consciousness, everything felt incomplete. Despite the enormous progress of modern science, I felt something essential was missing, not in the data, not in the observations, but in the core understanding of why the universe behaves the way it does.

This book is the product of that calling.

I began this journey not as a physicist, not as a professor or a theorist, but as a thinker, a relentless questioner. The answers I sought didn't come from textbooks or journals. They came from solitude, observation, logic, and faith. In 2023, after years of pondering, I articulated what I now call the Universal Law, a principle that, I believe, explains the birth, structure, and continuous evolution of everything in existence.

Writing this book is not about fame, recognition, or monetary gain. It is about contributions. It is my humble offering to my country and the world. A path toward infinite energy, space exploration, advanced healing, and perhaps most importantly, understanding the mystery of life itself.

I am aware that I do not have the complete scientific training to develop this theory in all its technical depth. But I believe the foundation I've laid is strong enough for others to build upon. I offer it openly to India's scientists, engineers, and visionaries, trusting that they will take this work forward with the same purpose and honesty that birthed it.

What follows in these pages is my journey, both scientific and personal. It is the truth as I discovered it. May it spark the courage in others to question deeply, think freely, and create fearlessly.

- **Akshay Vijay Gaikwad, Mumbai, India**

Introduction

For generations, the universe's greatest mysteries were cloaked in words like ***dark energy*** and ***black holes***. They were placeholders for what we could not explain, the ominous names for forces and structures that defied understanding. We accepted the darkness because we had no light.

But darkness was never the truth. It was the limit of our vision.

This book is about crossing that limit. About seeing what the universe has been whispering all along, not in riddles, but in patterns, motion, and balance too simple and elegant to stay hidden forever.

At the heart of this revelation is ***the Universal Law***, a principle that explains all unexplained phenomena in our cosmos. It sheds light on age-old paradoxes: why black holes behave as they do, why energy seems to emerge from nowhere, why time flows as it does, why galaxies spiral outward, and why life itself arises and transforms.

For centuries, black holes stood as puzzles, the cosmic prisons from which nothing could escape. But through the Universal Law, they gain a new meaning. Not as destroyers, but as essential defects within the space-time fabric. What we once called *dark energy*, a force we never truly understood, is revealed here not as a mystery to fear, but a natural expression of balance and resistance in the universe.

This law does not merely describe, it answers. The *how*, the *when*, and the *what* that have riddled mankind for centuries are no longer cloaked in uncertainty. The Universal Law reveals the reason for their existence and the principles of their working, uniting them into a single framework.

What I share here is not rebellion against science, nor rejection of the giants who came before. Newton sensed it. Einstein searched for it. Hawking glimpsed it. They were all reaching toward the same horizon, the law that ties it all together.

This is not philosophy, and it is not fiction. It is what the universe shows us when we dare to look without fear, when we replace mystery with clarity, when we discover that what was once dark is not so dark anymore.

If you've ever wondered why the cosmos behaves the way it does, or how life and death are not opposites but transformations within a larger order, this book is your invitation. Not just to read, but to understand. Not just to understand, but to awaken.

The darkness is lifting. The story of the universe is ready to be told in light.

Chapter 1: The Awakening

As a child, I remember staring at the night sky with wonder, not merely in awe of its beauty, but in confusion.

Why did it all exist?

What held the stars in place?

Why didn't they fall?

Where did the light come from?

Questions like these stirred something inside me long before I knew the language of science or the equations of motion.

But much later, as I began reading about modern physics, I encountered an unexpected feeling, not clarity, but disappointment. The deeper I went into textbooks, the more I realized the answers felt fragmented, isolated, or worse, circular. They described ***how***, but rarely ***why***. They offered math, but not meaning.

I was told to accept constants without origin, forces without cause, and phenomena without purpose.

At first, I thought the gap was in me. Perhaps I lacked the academic background, discipline, or patience. But over time, the feeling only grew stronger. Slowly, I began to suspect something profound: perhaps the framework itself was incomplete. Perhaps our models were like scaffolding for a building we could never see, not because it was hidden, but because we were not looking at the foundation.

It was not arrogance that led me to question the world's greatest scientists. It was a necessity. I needed to make sense of the universe in a way that did not ignore motion, did not rely on invisible constructs, and did not separate life from physics. I needed a principle that could explain not just equations, but existence itself.

And slowly, it came. Not as a sudden flash, but like dawn, steadily erasing the night.

I began to see that what we once labeled ***dark energy***, that elusive force accelerating galaxies apart, was never truly dark. It was a placeholder, a name born of ignorance, waiting for a deeper explanation.

At the same time, ***black holes***, long treated as paradoxes that defied understanding, began to take on a new meaning. They were not infinite abysses, but structured transformations. Not ends, but transitions. Not cosmic prisons, but part of the very architecture of balance in motion.

This was the awakening. Not to speculation, but to a truth I could no longer ignore: the universe was not fractured into mysteries, but bound by a single, underlying reality.

It became clear that gravity, time, energy, and even consciousness were not separate riddles, but outcomes of one connected order. By unveiling this order, what we once called black holes and dark energy could finally be understood, not as paradoxes, but as natural expressions of the same principle.

From that point forward, the night sky was never the same. I was no longer just a curious observer. I had become a witness to the unveiling of what had been hidden, and a messenger of what is now ready to be seen.

The darkness we once accepted is ***not so dark anymore.***

Chapter 2: The Limits of Modern Science

For centuries, science has been humanity's torchbearer, illuminating the darkness, explaining the unexplainable, and building the world we live in today. It gave us electricity, medicine, spaceflight, and computers. It gave us answers that, for the most part, worked.

But somewhere along the way, science became comfortable. It grew satisfied with fragmented models, incomplete equations, and constants we could not explain. It taught us to solve problems without asking if we truly understood them. It became more about predicting outcomes than grasping the deeper structure of existence. Instead of seeking truth, much of modern physics began to revolve around fitting observations into existing frameworks, even when those frameworks no longer made sense.

We are told that gravity is space-time curvature, but not why mass bends space-time at all. We are taught that time slows near massive objects, but not what time actually is. We are shown equations filled with constants like Planck's constant, the speed of light, and the gravitational constant, yet never offered insight into their origin. The Standard Model explains how particles interact, but not what gives them mass, or why they even exist.

And perhaps most alarmingly, the more advanced our theories became, the more detached they grew from human understanding. String theory, for example, operates in ten or eleven dimensions we cannot see, with vibrating entities that no one can detect. Quantum mechanics works astonishingly well mathematically, but no one can agree on what it means.

Are particles real?

Is the universe deterministic?

What is a wave function?

Are we just supposed to accept these mysteries as unknowable?

I could not.

Science should not be about memorizing outcomes; it should be about understanding origins. We need frameworks that begin with reality, not with mathematical abstraction. We need principles rooted in the nature of nature itself. We need a language that does not split the universe into separate forces and particles, but sees it as a continuous, dynamic whole.

That is what led me to seek an alternative. To look beneath the surface of all these models. Not to add another theory, but to uncover an underlying reality that made the existing frameworks appear as partial reflections of a much deeper whole. I did not start with the goal of disproving anything.

I started with a simple question:
What if everything we observe is the result of a single, universal tendency, one that governs matter, energy, and force alike?

That question became my compass. The more I followed it, the more I saw the cracks in our current understanding, not as failures of science, but as signs that something deeper was waiting to be revealed.

I remember watching a documentary by Dr. Michio Kaku. He explained the forces of nature and mentioned that if humans could find the relationship between them, all mysteries would be solved. That idea struck a chord within me. Later, I learned that Einstein, too, had spent much of his life searching for such a unification, seeking a framework that could link gravity, electromagnetism, and quantum phenomena.

In that moment, my pursuit of truth became more than curiosity. It became a mission. The questions I had carried for years were not mine alone. Great minds had wrestled with them before me. Yet while many had theorized, few had questioned the foundations themselves with the depth I now felt was necessary.

By then, I had already experienced something I could not unsee, a shift in perspective I described as an awakening. It was not the birth of a theory, but the recognition of a truth I could no longer ignore. A way of seeing the universe that revealed gravity, time, energy, and even consciousness as outcomes of one unified phenomenon.

And it was here that the one tendency showed its greatest power. It illuminated the darkest corners of human knowledge. What we had labeled ***dark energy***, and what we had treated as the paradox of ***black holes***, were not impenetrable mysteries after all. They were the cracks in the wall, the places where our incomplete frameworks faltered, and where the deeper law shone through most clearly.

Once seen, it changed everything. I could no longer be content with fragments. The one tendency was not just another principle; it was the light falling on the most shadowed parts of our understanding. It explained why galaxies expand, why matter collapses into black holes, and why the universe behaves as it does.

The question that remained was not whether this tendency existed, but how it shaped everything we see. That is where the journey led me next.

Chapter 3: The Missing Link

After questioning the foundations of modern science, I began searching more intentionally for what might have been overlooked. Not just a flaw in equations or an exception to a rule, but a fundamental piece that never made it into the puzzle in the first place.

What I found was not new. It was something old, an unanswered question that many great minds had chased, but never pinned down.

Einstein spent the latter part of his life searching for what he called a ***unified field theory***, a single explanation that could connect gravity and electromagnetism, the two major forces known in his time. He never found it. He died with that pursuit incomplete, the chalk still in his hand, the final line on the board unwritten.

Others came after him. Stephen Hawking, with his pursuit of quantum gravity. Richard Feynman, with quantum electrodynamics. Abdus Salam, Steven Weinberg, and Sheldon Glashow with electroweak unification. They all made remarkable progress, but still, something remained out of reach. None could say what gave rise to the force itself. None could say what space and time actually are.

Today, we still do not know why gravity exists. We can calculate its effects, but not its cause. We do not know what time is, beyond the way we measure it. We do not understand inertia, why a mass resists acceleration. We do not know why constants like the speed of light or Planck's constant have the values they do. And we certainly do not know why particles have mass, or why anything exists at all.

These are not minor details. They are the foundations of all science. And yet, science continues to operate with workarounds, equations that predict outcomes without explaining origins. Adding more dimensions, more forces, more particles, just to keep the current models afloat.

So, **what is the missing link?**

Why did so many brilliant scientists fall short?

Because they were building within a broken framework. They were trying to explain the whole without questioning the origin. They were mapping the behavior of systems without asking what makes a system exist in the first place.

At some point, I stopped searching for answers in equations and started searching in logic. I asked,

What is common to all physical phenomena, to all motion, all force, all energy?

Something was missing, not in complexity, but in simplicity. A root. A starting point.

Not another theory, but a principle.

The kind of principle that does not need ten dimensions to hold together. One that works because it is true, not because it fits.

That principle, I came to realize, was not buried in math or waiting in a particle collider. It was hidden in plain sight. Overlooked, not because it was too small, but because it was too fundamental.

And perhaps the reason I could see it, when others could not, was because I did not come from the traditional scientific mold. I was not trained to follow the path of observe, measure, infer, and publish. I did not know which questions were considered impossible, or which concepts had already been declared untouchable by consensus. I did not carry the weight of the constants or the caution of careers built on them.

I entered with nothing but a blank canvas and a burning desire to understand.

I was a dreamer. I relied on imagination more than instruction. In my mind, I could reach distances no telescope could see, travel faster than any particle, and enter regions where no data existed. I did not conduct experiments in labs; I performed them in thought, building one upon another, following only the trail of logic and possibility.

And in doing so, I stumbled upon ***the missing link.***

Piece by piece, it revealed itself. Not as another theory, but as a universal tendency, a principle woven through every phenomenon, the resistance within motion, the balance that shapes force, the root behind energy itself. It was present in galaxies expanding, in black holes collapsing, in the flow of time, and even in the energy we once called dark.

I would come to call it **The Universal Law**. Not something invented, but something uncovered.

And once seen, it changes everything.

Chapter 4: Motion Out of Darkness

When formal explanations failed me, I turned inward. I began conducting what I call thought experiments, not on paper, but in the mind. I would spend hours, sometimes entire nights, visualizing the universe not through equations, but through motion, space, and energy.

One of the earliest and most profound thought experiments came from something simple, a fragment of imagination inspired by a movie, ***Honey, I Blew Up the Kid*.** I imagined myself growing bigger and bigger. At first, the Earth was the size of a soccer ball in my hand. Then, as I expanded further, the Milky Way Galaxy itself appeared no larger than a Frisbee resting on my palm.

At that scale, a peculiar question arose. If I wanted to scratch my nose, would the signal from my brain travel to my arm at the speed of light, taking hours or days to arrive? Or would it be instantaneous?

The answer, I felt, was instantaneous. Because size is relative.

What if I could grow so large that the entire universe was a single object resting in my palm? Beyond it, perhaps in some unimaginable layer of existence, there might be beings for whom our universe is no more significant than a photon. This thought did not just challenge perception. It began to break the rigid limits that science had placed on the speed of light. It was the first crack in the darkness.

As my imagined size increased, so did everything within me. My neurons, my synapses, the dendrites and axons that carry thought, they too would scale up. The particles transmitting those signals would also be proportionally larger. In that frame of reference, vast distances would shrink. A kilometer might seem no different than a micrometer. The time for signals to travel would collapse, not because they exceeded light speed, but because scale itself had shifted.

In that frame, what felt instantaneous was not violating physical law. It revealed instead that laws like the speed of light may only appear fixed because we are fixed in size. When everything expands, perception, internal systems, and the space between objects expand with it. Speed and scale merge, and even what we once considered a universal limit begins to dissolve.

This led me to imagine what lies beyond the universe. Not other galaxies, not multiverses, but a truly empty region. A vast expanse without matter, without light, without anything measurable. Our universe, finite and bounded, is suspended in a sea of infinite emptiness.

That mental image haunted me because it made sense in a way no textbook explanation ever had. It made me see that the universe is not infinite. Only the emptiness beyond it is.

It was during one of these thought experiments, when I imagined myself holding the universe in my palm, surrounded by infinite nothingness, that I remembered Newton's first law of motion:

Every body continues in its state of rest or of uniform motion unless acted upon by an external unbalanced force.

As I visualized the universe suspended in that endless void, it struck me: there was nothing outside it. No other body, no external force, nothing beyond it to push or pull. According to Newton's principle, the universe could either be stationary or in uniform linear motion.

Why could the universe itself not be moving?

What if its entire structure, every phenomenon we observe, is part of a much larger, unified motion?

That was the striking moment, the point where I began to question the most fundamental assumptions I had been taught.

I examined celestial behavior more closely. Every object in space, planets, moons, stars, even galaxies, is always in motion, orbiting a larger parent body. The only time I ever saw stillness was when an object was physically attached to a larger one, bound by contact. The pattern was universal. Motion was the rule, stillness the rare exception.

Returning to my image of the universe in infinite darkness, I realized: since the universe touched nothing larger than itself, and nothing existed outside it to hold it in place, it could not be at rest. And since there was no greater body to orbit, it could not revolve either. Therefore, **according to Newton's law, the only possibility left was that the universe itself is moving in uniform linear motion.**

That realization was more than logical. It felt undeniable.

From there, the thought deepened. If the universe moves, it must displace the nothingness around it.

I imagined it like a submarine moving through an ocean. As the submarine glides forward, the water in front is pushed aside, opening a gap, and the water behind rushes in to close it. The submarine does not tear the ocean apart; rather, the ocean bends and flows to restore its own continuity. The resistance of the water gives the submarine both drag and buoyancy, but it also ensures the ocean never breaks apart.

In the same way, as the universe moves, emptiness itself must shift, rejoin, and resist. The fabric in front is displaced, and as the universe advances, that fabric rejoins behind. This rhythm of tearing and restoration gives rise to the bending of space-time that we observe. It is not destruction, but healing. It is the universe's way of moving while keeping continuity intact.

That was the light bulb moment. This emptiness, though made of nothing, behaved like a medium. It responded to motion. It resisted sudden change. It bent and restored balance.

The ultimate space-time fabric has one defining property: it always seeks to remain whole, continuous, and free. This tendency, though invisible, is absolute. And any fabric born from it carries the same inheritance.

Emptiness was not nothing. It was a structured, three-dimensional fabric, invisible, intangible, yet real. It resisted fragmentation and restored continuity. It was the silent stage upon which the universe moves.

This made me return to Einstein. He was correct: space-time does bend under the influence of mass. But what remained hidden was the mechanism. It is not the mere presence of matter that bends space-time, but its motion. As matter advances, it tears the fabric in front, and as it passes through, the fabric rejoins behind. This tearing and rejoining creates the curvature we perceive as bending.

This tendency of the fabric cannot be seen directly, but it can be felt. It is why matter does not disintegrate into the emptiness beyond, but remains bound together, traveling as groups, systems, and galaxies. It is the unseen stitching of the universe, holding all things in motion without letting them scatter apart.

In that moment, the mystery grew lighter. The darkness that once cloaked space-time revealed its hidden order. Motion was not an accident. It was the foundation of existence. And in that foundation, the first true law was waiting to be uncovered.

Chapter 5: Blackhole Mystery

Having now understood that the universe itself is a moving system, balanced delicately within a structured space-time fabric, I began to wonder: what happens when that balance breaks? What happens when a system loses its motion, its alignment? The answer, I discovered, lies in one of the most enigmatic features of the cosmos: the black hole.

Black holes have long stood as one of the most mysterious phenomena in modern physics. From Einstein's predictions to Hawking's radiation equations, black holes have remained a subject of awe, speculation, and paradox.

We are told that black holes are regions where gravity becomes so intense that not even light can escape. That space bends inward to a singularity. That time stops. But I always wondered:

Why should gravity behave differently here?

Why should space suddenly collapse completely?

And why does light simply disappear?

The traditional answers rely heavily on math and imagination but often fail to provide a deeper reason. They predict what might happen, but they don't explain why it happens. Something always felt incomplete.

As I continued exploring the patterns in space-time, one possibility started to take shape, not one based on extreme conditions or speculative singularities, but on continuity, motion, and the structure of the space-time fabric itself.

I began to sense a hidden principle, a tendency present in everything we observe, that might be guiding all systems toward a certain state of balance. A force not of attraction, but of resistance. A rhythm not of randomness, but of purpose.

We now understand that the universe is not still. It is in uniform linear motion, moving as one whole entity through the **Ultimate space-time fabric**, the infinite, undisturbed field that resists fragmentation. And within this universe is a second, distinct fabric, a fragment of the ultimate one that has been disintegrated and captured during motion. We've called this **the Universal Space-time Fabric.**

To visualize how these fabrics behave, I imagined a car. When the car is at rest, the air inside and outside is indistinguishable. Similarly, when the universe is at rest, or viewed without motion, the universal and ultimate space-time fabrics seem identical.

But when the car begins to move at high speed, a distinction forms. The air inside the car moves with it. A fly inside the car doesn't need to exert any effort to match the car's speed; it is simply carried forward by the medium it resides in. This is how matter behaves inside the universe. It does not need to create force to match the universe's velocity; it moves effortlessly with the Universal Space-time Fabric.

Because the ultimate fabric resists interruption, every disintegration of it, including the universal fabric, carries the same tendency to restore its free, unbroken form.

I then began to suspect that maybe, just maybe, black holes weren't endings, but returns. **Gateways, not prisons.**

Then I imagined something else: a hole in the rear windshield of the car. If the fly drifts into that hole, it is no longer protected by the interior air. It is left behind as the car moves forward. It doesn't fall because of gravity; it simply loses its place in the motion of the system.

To someone still inside the car (or inside the universe), the fly appears to fall backward. But it's not being pulled; it's simply left behind as we surge forward. This is how black holes appear to us: not as attractors, but as points where something lost its place in the motion.

This image struck me deeply. Could black holes be the same?

What if a black hole is not a region of unimaginable gravitational pull, but a defect in the Universal Space-time Fabric? A place where the moving fabric tears or reconnects with the ultimate fabric. A region where matter, like the fly, simply loses its connection with the universal motion and is left behind.

In this view, a black hole is not pulling matter inward. Rather, matter entering that region ceases to move with the universe. It is no longer part of the system in motion. From our perspective, embedded in that moving system, it appears to disappear. But it is not falling. It is being left behind.

Interestingly, observational data never truly confirms a dense core; we only see the effect: missing matter, frozen light, and curving motion.

This also provides an elegant explanation for one of the most puzzling aspects of black holes: why we cannot see light escape them. If the velocity of the universe is greater than the speed of light, then anything that stops moving with the universe, including light, would never be seen again. Not because it is being pulled inward, but because we have already moved ahead of it.

Since the universe is in motion through a medium, and black holes prevent light from escaping not because of compression but because of loss of motion, it implies that the background motion of the universe exceeds the speed of light. This isn't a violation of relativity, but a redefinition of context: light travels within the universal space, but the universal space itself moves faster than light relative to the ultimate space-time fabric.

It also reveals something more fundamental: the black hole is not a dense object at all. It is a threshold. A location in space where the Universal Space-time Fabric breaks and reconnects with the Ultimate Fabric. In this region, motion ends, and unity begins.

And what happens to matter that enters this region? Since it is no longer in motion, it no longer disrupts space-time. It no longer creates structure. It no longer exists as an independent entity. It dissolves back into the Ultimate Space-time Fabric, like a drop of water returning to the ocean. It becomes one with the whole.

This led me to a powerful realization:

1. **The velocity of the universe must be greater than the speed of light, or we would see what happens inside a black hole.**

2. **Black holes are not high-gravity zones, but defects in the space-time fabric that reveal the boundary between structured motion and infinite unity.**

They are not monsters hiding in the dark. They are clues that the fabric of space is real, structured, and part of something larger. Clues that motion is not a local effect, but a universal one. Clues that the universe itself is in motion through a medium that holds the secret to everything.

And so, what once stood as the darkest mystery in the cosmos is **not so dark anymore.**

In the next chapter, we will explore the force that is present everywhere, the one we all feel, and yet have not explained fully. Its effects have been described with remarkable accuracy by mathematics, but its very origin remains elusive. This is the force we know as gravity.

Chapter 6: The Restoring Force

We've journeyed from understanding motion itself to witnessing how it might give rise to structure. But something else must also be at work behind all these processes, something subtle, yet everywhere. Could that something be gravity?

We're told that gravity is a force of attraction between two masses. The more massive an object, the more gravity it exerts. But this idea, while practical, has always seemed incomplete.

Why should mass have the power to attract?

What gives it this mysterious influence?

Perhaps we've misunderstood gravity, not as a product of mass, but as a consequence of motion through the medium it disturbs: **space-time itself.**

If motion can fracture the undisturbed ultimate space-time fabric and create a new internal one, as we saw in the last chapter, then that internal fabric must still inherit the same property to preserve its free, continuous form. So, what happens when this internal fabric is disturbed again by moving matter within the universe?

When the universe began its motion through the ultimate space-time fabric, it fractured the undisturbed medium, creating a moving internal fabric, the universal space-time fabric. This new fabric, though a byproduct of motion, may still carry the fundamental tendency of the ultimate fabric: to remain continuous, uninterrupted, and free from deformation.

As groups of matter begin to move and cluster inside the universe, they each trap a portion of the universal space-time fabric around them. These trapped portions are distortions, imperfections in the smooth flowing form of space-time.

If the fabric wants to remain one undisturbed 3-dimensional entity, wouldn't it push back against this disruption?

Imagine a submarine submerged in the ocean again. As it moves forward, it displaces the ocean water at the front, and that water rejoins smoothly behind it. This is much like how the universe, moving through the ultimate space-time fabric, creates a flow, disintegrating the fabric in front and allowing it to reunite at the rear.

Now, take this further with the car analogy. A moving car creates a defined space within itself, separated from the air outside. Likewise, a moving system within the universe creates its own internal space-time fabric, distinct from the undisturbed fabric beyond. As clusters of matter move through this universal fabric, they trap and shape it, breaking its uniformity and forming local disruptions that the fabric naturally resists. This disruption, and the resistance to it, is what we experience as gravity.

Now, take the analogy of a car in a wind tunnel. The wind wraps tightly around the surface of the car, flowing smoothly along its shape. Replace this wind with the space-time fabric and the car with a body in motion inside the universe. The fabric flows with the universe and bends around every object within it. Introduce a smaller object near the car, and what happens? The flow must bend around this object too, engulfing it and rejoining behind it. And because the space-time fabric wants to maintain its one continuous flow, it begins pushing the smaller object toward the larger one, to erase the disturbance the second object caused.

This, then, is gravity.

The force exerted by the space-time fabric to maintain its one 3-dimensional continuity when matter interrupts it by virtue of its motion is called gravitational force.

It's not the matter that is pulling other objects in. It's the space-time fabric pushing them together in its attempt to restore continuity. The more mass an object has, the more it disturbs the flow of space-time. The stronger the disturbance, the stronger the restorative push.

If an object were completely at rest, it would not disintegrate space-time. No flow would be generated, no resistance would form, and there would be no gravity. This may explain why we see no resting objects in the cosmos. Without motion, there would be no force binding particles into structures. No motion, no space-time flow. No flow-no gravity. No gravity-no mass.

This means gravity arises not from mass itself, but from the motion of mass through space-time, and the pushback of space-time to remain whole.

Black holes, then, are not monsters hiding in the dark. They are regions where the resistance of the fabric has collapsed entirely under extreme fragmentation. In these zones, matter enters the ultimate space-time fabric and loses its motion with the universe. The ultimate fabric begins to engulf and dissolve this matter, deforming itself by its presence, just like the hole in the rear windscreen from our past thought experiment, where the free, flowing medium outside begins to seep in, erasing the boundary and reclaiming continuity.

This reinforces what we discovered in the last chapter: that black holes are not exotic exceptions, but natural results of how space-time reacts to intense disruption.

In this view, gravity is not a pulling force from mass but a restoring force from space-time. A memory of motion, a protest against fragmentation, and a tendency to heal.

The familiar saying that "mass tells space-time how to curve, and space-time tells mass how to move" takes on a deeper meaning. Mass is not the cause but the symptom. The curvature is not a product but a protest. Motion causes disturbance. Disturbance bends space-time. Space-time responds.

Gravity, then, is not a property of mass; it is a memory of motion.

Getting trapped in an orbit around a larger body is, in fact, the clearest sign that a group of matter has reached a temporary state of equilibrium with the space-time fabric surrounding it. This stable, balanced motion echoes the larger equilibrium we believe the universe itself has achieved.

In the next chapter, we will look more closely at how these groups of matter, each locked in a unique motion and surrounded by a portion of trapped space-time, begin forming organized structures. This will be the beginning of our discussion on what we later define as systems of matter, the hierarchical building blocks of the cosmos.

Chapter 7: Systems of Matter

Now that we have seen how gravity emerges from space-time's resistance to deformation, a deeper question arises:

Why do different systems behave so differently?

Why does a planet orbit while a galaxy spirals?

The answer lies in the architecture of systems. Each inherits a slice of motion from its parent and tries to balance it. To see this clearly, we must explore how systems trap space and time and how that shapes their behavior.

After redefining gravity not as a mysterious pull but as a reaction of space-time to motion, another question emerged. If the universe itself is in motion, then what about everything inside it? Does matter simply move freely within this grand motion, or does it create its own? Does it trap its own slice of space-time and leave its own imprint?

To explore this, I returned to the moving car analogy. A car traveling along a highway creates a distinct internal environment. Within it, everything shares the car's velocity. A fly inside does not struggle to match the speed; it simply moves with the air around it.

This is much like the universe, a system moving through the ultimate space-time fabric. The motion of the universe creates its own internal fabric, the universal space-time fabric, within which everything else exists.

But now imagine a balloon floating inside the car. Unlike the fly, it moves relative to the car, creating its own smaller environment. This is precisely what happens in the cosmos. Inside the universe, galaxies move not at the velocity of the universe itself, but relative to it. That relative motion traps a portion of the universal fabric. These are what we now call **Galactic Systems.**

Each galactic system comprises two elements: the matter it contains and the slice of fabric it traps through motion. Together they move as one through the universal fabric, and the gravity within the galaxy arises from its own fabric, striving to preserve continuity.

Within galaxies, stars orbit and rotate, trapping portions of galactic fabric to form **Stellar Systems.** Planets revolving around stars trap stellar fabric to form **Planetary Systems.** Moons orbiting planets, or satellites around Earth, trap planetary fabric to form **Satellite Systems and Man-Made Systems.** Each system is a distinct space-time structure, moving within its parent field and generating its own influence.

These systems are not isolated. They form a hierarchy of nested motions, each smaller system born within its parent. Each carries a trapped space time fabric taken from its parent, and each is striving to return and become whole again with the ultimate space time fabric.

But just like the air inside a moving car cannot rejoin the air outside until the car stops, the trapped fabric of a system cannot merge back with its parent fabric while the system remains in motion. Only when the motion ends can the fabric reunite and restore wholeness.

This was another light bulb moment. The disintegration of space-time, caused by the motion of a system, is never permanent. It is always pressing to merge back with its parent, always seeking continuity. But it can do so only when the system's motion ceases.

At the very base of this hierarchy lie **Quantum Systems.** Fundamental particles like photons and electrons are the smallest systems of matter. They were likely the first to form, long before galaxies or stars. Traveling at extremely high velocities, often near the speed of light, they approach the equilibrium of the universal system more closely than any other.

And yet a further question arose. If these particles were truly free within the universal fabric, would they not behave like tiny galaxies, moving on their own within the flow? In practice, they are not free. They live inside larger systems, within planetary fields and stellar neighborhoods. The space-time fabric of these parent systems surrounds them and sets the conditions for their motion.

Although these particles trapped the universal space-time fabric directly, their small scale makes them sensitive to the gravitational field produced by the planetary or stellar fabric they inhabit. Under that influence, they cling to one another. Their individual freedom of motion ceases, and they gather into larger bodies inside the parent system. In this way, they become the material of planets and moons, and stones, shaped by the parent fabric while still carrying a slice of the universal one within them.

The hierarchy becomes clear:

Universal System

Galactic Systems and Quantum Systems

Stellar Systems

Planetary Systems

Satellite Systems

Man-Made Systems

Galactic and quantum systems share a similar place. Both are early products of the universe's development. Both show strong gravitational and fabric behavior. Every system exists because of motion relative to something larger. And that motion always gives rise to its own fabric, its own field, its own presence.

On vast scales, galaxies dominate. On the smallest scales, quantum systems echo the same laws. Complexity does not come from size but from position in the hierarchy of space-time.

Man-made systems are no exception. Satellites, spacecraft, and even machines on Earth all move within the planetary fabric and trap portions of it as they accelerate. They, too, are systems of matter, small but obedient to the same law. Even our creations cannot escape the rules of space-time.

And here lies the most striking insight. Of all the systems in this hierarchy, only the Universal System has reached perfect equilibrium. It moves in a straight line at uniform velocity, undisturbed and free. All other systems are still striving. Galaxies, stars, and planets move, but their motion is curved, rotational, and restrained. They orbit because they cannot yet achieve the freedom of their parents' straight-line motion.

Orbits, then, are not failures. They are solutions. Signs of systems finding the best balance possible within the constraints of their parents' fabric. Stable. Repeating. Enduring.

Yet even with this clarity, a deeper question remained. We now understand systems and their hierarchy, but not the most profound mystery of all:

How did matter itself arise?

How did trapped motion become mass?

And why does this mass generate gravity?

These questions stood as the last shadows at the edge of thought. But they would not remain shadows for long.

In the next chapter, we uncover the hidden drive that turns energy into matter, the universal effort to reach unity. This is where we arrive at the principle behind all principles, the only energy present in every system. What we once called dark energy, but which **is not dark anymore.**

Chapter 8: The Ultimate Velocity

Having uncovered the systems of matter, let us turn to the universal system itself. What does it mean to imagine the entire universe in **uniform linear motion?**

If no external force acts on it, and if there is no acceleration and no deceleration, then no work is being done on the universal system. By definition, there is no force being applied to it. And yet a question persists.

If the universe is moving, must it not possess kinetic energy by virtue of that motion?

If so, where does that energy come from?

Earlier, the behavior of black holes gave us a demanding clue. The velocity of the universe must be higher than the speed of light. If that is true, the ultimate velocity is very high indeed. But the law of conservation of energy does not bend for intuition. Energy is not created and not destroyed. It flows from one form to another. The kinetic energy of the universal system must therefore be balanced by the energies that exist within the universe itself.

What are these energies?

There is the energy associated with mass itself. We owe a clear debt to Einstein, who demonstrated that rest energy and mass are connected by a simple and profound relationship. Let us denote rest energy as R.E. Then

R.E = m c^2 (1)

Even when the position does not change, mass carries energy. When portions of that store are released, they appear as light, heat, sound, and other familiar forms we measure every day.

Those familiar forms are not a separate ledger beside mass. They are exchanges with the same store. They move into and out of rest energy as conditions change. To avoid double-counting, it is honest to treat them as expressions of R.E within the total account.

There is also an earlier reserve to consider, a reserve that must have existed for the universe to reach its **present motion**. It is the unspent drive of a system to reach the ultimate velocity of its own fabric. When the system is free, this reserve becomes motion. When the system is restrained, the reserve persists. With sustained restraint, it settles into mass. We once called it dark energy because it was unseen. We will keep that name here and write it as D.E.

With these meanings in place, the balance for the universal system can be set down without hurry. The total internal energy should equal the kinetic energy of its motion as a whole:

K.E = R.E + D.E (2)

If the universal system is now in uniform motion, the phase of acceleration is complete. The reserve that was needed to reach this state has already been used. For the universe at present, this suggests

D.E = 0 (3)

Substituting (3) into (2) gives

$K.E = R.E$ (4)

What speed must satisfy that equality? Let mmm denote the effective total mass of all contained systems in the universe at the present epoch (not the mass of the universal system itself), and let vvv be the speed of the universal system. Then the standard forms are

$K.E = (1/2)\, m\, v^2$ (5)

$R.E = m\, c^2$ (6)

Equating (5) and (6) using (4) gives

$(1/2)\, m\, v^2 = m\, c^2$ (7)

Normalize both sides of (7) by the common scalar m with $m \neq 0$, which yields

$(1/2)\, v^2 = c^2$ (8)

Solving for vvv gives

$v = \sqrt{2} \times c$ (9)

With c=299,792,458c = 299,792,458c=299,792,458 m/s, the numerical value is

$v \approx 4.23970560 \times 10^8$ m/s (10)

Set down this simply, and the meaning becomes clear. For the universe, the reserve we call D.E has already been spent, and the motion we observe is its quiet result. Yet the story does not end there. Other systems do not live in this finished state. Their ledgers are still open. They carry motion, they carry R.E, and they still hold a reserve that has not yet found its way into speed.

What exactly is that reserve inside a galaxy, inside a star, inside a stone in our hand? How does it move? How does it trade places with R.E and K.E as the fabric eases or tightens? These are not footnotes. They are the heart of the question.

In the next chapter, we will speak plainly about the Universal Law, not as an announcement but as the most natural reading of this ledger. We will follow how K.E, R.E, and D.E shape the behavior of every system, and why the reserve that once seemed hidden is, in truth, only waiting to be seen. The darkness in its name begins to lift. It is not so dark anymore.

Chapter 9: The Universal Law

After discovering the value of the ultimate velocity of our universe, our attention must now turn to what all the revelations so far truly mean. The property of space-time, the presence of dark energy, the birth of mass, and the behavior of systems of matter **all point toward a deeper principle**. Together, they suggest a law, one that is not separate from known physics but one that reveals the common tendency behind all energies.

We have seen that motion is not a byproduct but the very essence. Every system holds within it a reserve of intent that we call dark energy, the urge to align with the velocity of its parent system. This striving is constant. It never disappears, even when hidden beneath form and structure.

When a system is free, dark energy becomes acceleration. When resisted by the parent's space-time fabric, unspent energy condenses into mass. Matter is not a separate substance. It is motion-restrained and held in place, the memory of what could not yet move.

From this contest comes a fragile balance. Dark energy pushes forward, space-time resists, and an orbit is born. In such an arrangement, no more mass is created and no more dark energy is lost. Yet the striving remains alive, ready to awaken at the smallest disturbance.

We can now state the law.

The Universal Law

All matter or systems of matter are in uniform linear motion with ultimate velocity unless acted upon by an external unbalanced force.

This law resembles Newton's first law, but it completes what Newton left unsaid. Newton described inertia but not the deeper striving of systems for their parents' motion, nor the resistance of space-time that holds them back. In this tug of war, galaxies, stars, planets, and even fundamental particles come into being. The universe is not a collection of still objects. It is the living balance between motion and restraint.

The pattern of ascent unfolds across a hierarchy. A particle seeks to free itself from the atom. If liberated, it aligns with the lattice. From there, systems rise to the scale of planets, stars, and galaxies, each time striving again for release from their parent. At each promotion, the same story repeats. Dark energy drives the attempt to escape, space-time resists, and equilibrium forms anew. The climb continues until nothing remains to resist, and the system aligns with the ultimate velocity of the universal system itself.

Equilibrium is never final. It is only a pause. A single disturbance can turn mass back into dark energy, reignite acceleration, and break the old bonds. In this way, every system carries within it the seeds of liberation and the memory of motion not yet fulfilled.

The Universal Law completes the thought Newton began. It holds his truth and widens the frame. All systems strive for uniform motion with ultimate velocity. Dark energy drives, space-time resists, and from their struggle comes mass, structure, and the very existence we see. Energy, mass, and motion are not separate but three expressions of the same pursuit.

The universe is not a place of rest. It is a place of restlessness. The Universal Law is the memory, the desire, and the path of all motion.

The Universal Law is not only about what systems do. It is about what they are trying to do. And from that effort, and from that resistance, all phenomena are born. It is the cosmic story of a single desire: to move freely, to complete the motion, to dissolve back into equilibrium.

With this insight, **I realized: systems do not form randomly.** They form in resistance to motion. Motion when restrained creates mass. Resistance gives birth to form. From galaxies to atoms, all are balancing their energy to reach the final motion.

There is no more darkness now, only light. And that light must be shed on the phenomena we see in our cosmos, so that we may understand them. The rest of this book is dedicated to witnessing the law in action, from black holes and quantum tunnels to life itself. The law is simple. The effects are infinite.

Chapter 10: The Birth of Our Universe

After uncovering the Universal Law, a new question arose. If every system strives toward equilibrium by reaching its ultimate velocity,

What happens when one of them succeeds?

What is the fate of a system, like a galaxy, that fulfills its cosmic destiny?

To answer, we must return to the hierarchy we have uncovered. Systems do not simply exist; they evolve. Quantum systems combine into atoms, atoms into planets, planets into solar systems, and solar systems into galaxies. Each step is a promotion, a system rising closer to the velocity of its parent. All are striving, all are trying to reach equilibrium.

Yet most are restrained. They become locked into orbits, balanced but incomplete. The space-time fabric of their parent system pushes back, resisting their acceleration. This restraint converts their dark energy into mass, holding them in a state of temporary equilibrium. But a rare few do not remain bound. They accelerate, break free, and edge ever closer to the ultimate velocity, the pure motion of the universal system itself.

Particularly for **galactic systems,** which sit directly within the universal fabric, there are three possible fates. Some lose motion and collapse into black holes, dissolving into the stillness of ultimate space. Others are caught in orbits around black holes, restrained by their intense gravity and kept in fragile equilibrium. These orbiting systems constantly balance dark energy and mass, converting one into the other in order to survive. But the rarest fate is escape. A galactic system that slips free continues its acceleration, aligning more and more with the velocity of the universal system itself. Only in such rare escapes does something extraordinary occur.

This led to a profound realization: there are two ways by which matter can fulfill its cosmic goal of reaching the ultimate velocity.

The first is the most dramatic. A system may accelerate so completely that it escapes the gravitational field of its parent universal system. At this threshold, the system's velocity is nearly indistinguishable from that of its parent. But what happens then? Does it dissolve into the parent's flow, its mass finally released back into pure motion? Must its restrained energy, all its stored mass, convert into radiant energy so that it can merge seamlessly into the undisturbed equilibrium of the ultimate space-time fabric?

If so, the law of conservation of energy demands an answer:

Where does this energy go?

The universal system is already in perfect motion; it cannot absorb more without breaking its equilibrium. The space-time fabric resists. But when resistance fails, when energy overwhelms the restoring capacity of the fabric, the system is expelled. And this expulsion is not chaos. It is creation. It is the birth of a new universal system. A universe born from an imbalance that could no longer be contained.

The second path is smaller in scale, but no less important. Even within a universal system, particles sometimes mirror this escape. In unstable nuclei, restrained motion can no longer hold. Dark energy is released. Mass dissolves into pure energy, igniting a cascade. We call this nuclear fission. At its root, it is the same truth: motion when restrained becomes mass, and when restraint lifts, mass dissolves back into motion.

Einstein captured this in a single equation: $E = mc^2$. In the Universal Law, this is not only a formula, but a deeper prophecy. Mass is motion in waiting. When restraint breaks, energy flows. The equation is not just equivalence; it is fulfillment. It is the return of stored motion to its purest form.

On cosmic scales, this release is not a chain reaction but a rupture. It may explain some of the most violent events we observe: gamma ray bursts, ultra-high-energy cosmic rays, unexplained flashes of brilliance. These are not anomalies. They are glimpses of matter nearing escape, pressing against the fabric of the universal system.

But when such eruptions become too powerful, the fabric tears. A black hole is born. Not as chaos, but as order restored. The black hole is the universe's pressure valve. Just as a steam engine vents excess pressure to survive, space-time vents its excess energy by opening a tear. A black hole is not infinite gravity; it is the point where motion collapses into stillness, where the fabric reconnects with equilibrium.

Think of a car racing in a wind tunnel. As its speed increases, the airflow strains to wrap around it. At some point, the flow ruptures. So too with matter and the space-time fabric. As systems accelerate, the fabric stretches to contain them. When the strain exceeds its limit, it tears. The tear is the black hole. The release is the birth of structure.

Thus, every newborn universe begins with such a tear. At its center lies a black hole, the exit point of its birth. Around it, unconverted matter organizes, caught in motion, spiraling into structure. Freed from the old restraint, the new system moves forward in uniform linear motion, carrying with it the memory of its parent but with velocity all its own.

And so, the process of universe formation is not a departure from the law, but its purest expression. Every explosion, every black hole, every system is simply the story of equilibrium pursued and resisted.

The universe is not rigid. It is alive. It balances motion and resistance at every scale. Systems evolve not only by force, but by memory, carrying forward the motion they inherited. From the smallest particle to the grandest galaxy, equilibrium is never still; it is a dance. And when one dancer moves too fast, the stage reshapes to absorb or redirect that motion.

In the next chapter, we will explore the foundations of this new universe. How the central black holes become gravitational anchors. How galactic systems find their place in orbit. And how space time itself begins to weave the cosmic architecture we now see.

This is not a tale of chaos. It is the story of cosmic memory. The story of restrained motion seeking release, of mass dissolving back into energy, of universes being born in the restless pursuit of equilibrium.

Chapter 11: Foundations of a New Universe

In the previous chapter, we explored how new universes can be born from systems that achieve ultimate velocity. To better grasp this concept, consider a powerful analogy: a uranium atom, the heaviest naturally occurring atom. It constantly emits alpha and beta particles, losing mass in high-energy bursts. This is not chaos; it is a structured release, a process toward stability. In the same way, galaxies may be escaping from our universal system, ejected fragments carrying stored energy and motion, like the alpha particles of a decaying nucleus.

If we could view our universe from above, perhaps it would resemble a cosmic atom, with galaxies being flung outward like charged fragments from its unstable core.

Interestingly, when I read about the modern scientific discovery that every galaxy appears to have a black hole at its center, it reaffirmed everything I had come to understand. This was not a surprising discovery for me; it was expected. I had already sensed that such anchoring centers must exist in all large-scale systems to stabilize the trapped space-time fabric in the system's motion and preserve the structure of their parents' space-time fabric. It was one more moment when I knew this Universal Law I had uncovered was not just a theory. It was real.

But what happens after that energetic birth?

When a new universe emerges, it does not erupt into disorder; it unfolds. The release of energy is of such high magnitude that it ruptures the space-time fabric within the trapped segment of the ultimate space-time fabric, creating a black hole. This tear acts like a sinkhole, rapidly folding the surrounding space-time into a dense core. The high energy that caused the rupture becomes concentrated here, and space-time, in trying to restore equilibrium, bends inward, giving birth to a black hole that becomes the gravitational anchor for the newborn universe.

All other **particles begin to form** and organize around this **central black hole, not randomly,** but guided by the direction and momentum inherited from their parent system. This motion remains uniform and linear, defined by the specific trajectory at which the parent system was ejected. Because of this unique directional inheritance, the new universal system travels with a velocity even higher than its parent and continues its straight-line motion undisturbed.

This does not imply faster-than-light travel within a closed system but rather reflects the clean momentum transfer within the undisturbed ultimate space, free from any opposing gravitational fields. The new system's velocity is measured relative to its point of separation, and motion continues uninhibited. This directional separation ensures that no two universes ever collide: each traces its own irreversible and isolated path through the ultimate space.

Everything this new universe contains, from the smallest particle to the largest system, stems from that first moment of motion, from the energy released by a departing galactic system. This newborn universe carries the tension, structure, and laws of the parent system but begins forming its own hierarchy, a layered evolution of motion.

The first to emerge are not galaxies or stars. They are the most basic systems: the quantum systems. These are not just particles; they are the smallest trapped motions, each capturing a fragment of the space-time fabric of this newly formed universe. Quantum systems are unique because they trap the universal space-time fabric directly at birth. Photons, electrons, quarks, they do not sit still. They move, spin, and vibrate. In fact, they cannot stop. That motion is the signature of their birth. It is the continuation of momentum from the larger system they emerged from. These quantum systems are not formed from nothing; they are carved from the energy released as a galactic system crossed the boundary of its parent universe and began to unfold. Later, when they bond under gravity and form larger objects, they fall into the hierarchy of systems, a point we will return to in detail in the quantum chapters ahead.

Although small, these quantum systems act just like galaxies or stars. They trap space-time. They are in motion. They possess internal equilibrium. They are the primary systems of this new universe.

Their motion is not random. It is part of the larger direction set by the universal system's birth. This collective direction defines the structure of the new space-time fabric, just as wind patterns define the motion of air or ocean currents guide waves. These tiny systems hold the memory of their birth, and that memory manifests as structured motion.

As quantum systems evolve and interact, their trapped space-time fabric produces localized gravitational effects. These subtle attractions and repulsions initiate the next level of structural complexity: first atoms, then molecular clouds, and eventually, celestial systems.

Just like in our own universe, galactic systems, solar systems, and planetary systems all start as clusters of motion. Each new layer is a promotion of the previous, built through motion, collision, and balance. Nothing in the new universe is truly at rest. Motion defines everything.

One thought experiment brings this into sharp focus. Imagine a standard model atom: a dense nucleus at the center with electrons locked in orbit. Now compare that to a solar system, a massive sun at the center, with planets revolving in stable orbits. If orbits are not stable, the planets drift, possibly falling into the gravity of another star. The structure of atoms is not a coincidence. The universe is layered, scaled, and mirrored. What we observe at one scale may echo in countless others.

Now, imagine particles within this new universe forming and colliding. One such collision creates a defect in the surrounding space-time fabric. This defect does not just sit passively; it begins to trap other systems in orbit. A black hole is born. Not as an anomaly, but as a structural necessity. It becomes the central anchor for this growing collection of systems.

This leads us to distinguish between types of systems. Primary and secondary systems, like universal and galactic systems, form around black holes. These black holes anchor grand-scale motion and stabilize entire space-time fabrics. Other systems, stellar and planetary, do not require black holes at their center. Their gravitational cohesion comes from the galactic or stellar space-time fabric they belong to. Their motion, though significant locally, does not deform the universal fabric the way galactic systems do.

If black holes arise from collisions and defects, and begin trapping systems, then every galaxy must contain a massive black hole at its center. This black hole provides the stability and binding force for all child systems, solar systems, planetary systems, and more. This structure is essential. It gives the galaxy cohesion and enables it to move as a unified body within the universal space-time fabric.

This realization raised a deeper question. If galaxies require a central black hole to bind their systems, a stabilizing point of space-time tension, could it be that smaller systems follow the same principle? Could it be that atoms, and even biological cells, have their own microscale equivalents, points of concentrated motion, tiny ports to deeper layers of the fabric?

We have not observed such micro black holes yet, but isn't it curious how matter always seems to organize around a center? And what if, at the scale of life, that center is something deeper?

something like the human heart?

Though speculative, it aligns with the law of motion. If structure emerges from trapped motion, and black holes are regions of concentrated deformation, then such stabilizing centers might exist even at microscales. Perhaps this is why we have not isolated the mechanism of consciousness or fully understood death.

I began to wonder, what if, within us, the human heart acts as such a port?

A central organizer, a black hole of sorts, around which the motion of life revolves. When its rhythm fails, the system collapses, and remaining motion dissolves into the ultimate space-time fabric. This could explain why death feels like disappearance, and why consciousness is so elusive to replicate.

This insight painted a chillingly clear picture. I recalled the film "Honey, I Blew Up the Kid," and a question resurfaced.

What if I grew so large that galaxies appeared to me as molecules, solar systems as atoms, and the subatomic particles inside them as particles in a collider?

What if our entire universe is like one electron in a copper atom, endlessly looping through a long wire that itself is just a structure within the ultimate space-time fabric?

Perhaps the universe is not just layered in space and time, but also in scale and perception. Particles may be galaxies on a different scale. Galaxies may be atoms in a larger structure.

It struck me that size may be the key to understanding the cosmos. If we lived on an electron, how could we grasp the full atom, let alone the entire molecule? Similarly, from our perspective, we may never fully grasp the complete structure of our universal system.

Yet, throughout all of this, the Universal Law continues to operate. Every system is in motion. Every system resists anything that restrains that motion. When restraint occurs, dark energy stores as mass. When freedom returns, mass converts to energy, accelerating the system again.

This cycle continues. Particles become atoms, atoms become molecules, and molecules become systems. And in rare, delicate places, the balance becomes just right.

Life begins.

On planets, dark energy achieves equilibrium by converting into the mass of matter. However,

What if some residual dark energy remains?

This leftover energy could convert directly into the cellular structures of life. Life, then, may not just arise from a chain of chemical reactions but could represent a direct conversion of dark energy into self-organizing, adaptive motion.

This connects all forms of energy, mass, motion, and life back to dark energy, the original source. Life is not a byproduct; it is a continuation of the universe’s push toward equilibrium through increasingly complex motion.

At first, it seems like a miracle. But with the Universal Law in mind, life is not a miracle. It is inevitable. It is what happens when motion gains complexity, self-organization, and memory. Life is motion that adapts.

Looking back, one truth stood out. A system giving birth to another through a burst of energy and structured unfolding mirrors what we call the Big Bang. We ask what gave birth to our universe, but the first motion that disturbed the ultimate space-time fabric, and the reason for that motion, may always remain unknown. Like a chain reaction, we trace the links, but never the origin.

This is the hidden truth of our universe. To preserve equilibrium and remain in existence, it must constantly regulate itself: by dissolving excess energy into black holes and by birthing new universal systems when necessary.

This Universal Law not only explains the structure and behavior of everything we see, but it also answers the age-old question:

What is the fate of our universe?

And do multiverses exist?

Yes, but not in the traditional way we imagine them. These are not parallel realities layered atop one another, but unique systems, born from energy overflow, forever traveling away from us in different directions and at their own velocity, never to collide, never to return.

With this picture of a new universe unfolding from motion, let us now step back and review what we have uncovered so far, before we apply this lens to the strangest phenomena of all: quantum behavior, life, and consciousness.

In the next chapter, we will pause to reflect. Before diving into the phenomena we aim to explain, we will briefly revisit the discoveries we have made so far, how space time, motion, dark energy, mass, black holes, and systems of matter fit together, to refresh our minds for the journey ahead.

The foundation is laid. Now the real decoding begins.

Chapter 12: A Journey So Far

Before we dive into the exploration of the strange and fascinating phenomena that exist in our cosmos, quantum oddities, the nature of life, and the mystery of consciousness, let us pause for a moment. We have traveled far, uncovered much, and built a new framework for understanding the universe.

To truly appreciate what comes next, we must take a breath and reflect. This chapter is not repetition, but crystallization. It is a recap of the key discoveries that form the spine of our journey, the pillars of the Universal Law.

1. The Motion of the Universe

We began with something bold yet beautifully simple: the universe is in motion. Not chaotic or circular, but a grand, uniform, linear motion through an undisturbed field we called the ultimate space-time fabric. This is the original stillness, the canvas upon which everything else unfolds.

This motion does not require an outside force. It is the residue of a past transformation. And as the universe moves, it creates its own field, a universal space-time fabric, a secondary layer that carries everything within it forward.

Only universal systems move in this perfect, straight, uniform way. They have exhausted all their dark energy and reached equilibrium with the ultimate space-time fabric. Every other system remains restrained.

2. The Birth of Systems

From this grand motion, systems emerge. They are not just physical structures, but events within space-time, formed from matter and the fabric it traps. A system begins when part of the parent system's motion becomes restrained. That unconverted motion is what we call dark energy.

It is important to restate that in this framework, dark energy is different from the version used in mainstream cosmology. Here, dark energy is the stored potential of restrained motion, not a mysterious force accelerating cosmic expansion, but the unfulfilled tendency of systems striving to reach their parent's ultimate velocity.

Galaxies, stars, planets, and atoms are all systems, nested within parents. Each inherits motion but also inherits imbalance. That restraint transforms dark energy into mass. Not because the system seeks mass, but because motion must go somewhere. When velocity is denied, structure is born.

3. The Universal Law

All motion follows a rule:

All matter or systems of matter are in uniform linear motion with ultimate velocity unless acted upon by an external unbalanced force.

This single sentence defines the behavior of everything we observe. It sharpens Newton's first law, but extends it by revealing the bigger picture: all systems strive for equilibrium, the ultimate velocity of their parent system.

If they are free, dark energy converts into motion. If they are restrained, dark energy is expressed as mass.

4. The Hierarchy of Systems

Every system, universal, galactic, stellar, planetary, manmade, or quantum, is defined not by size, but by placement in the hierarchy of motion and resistance. Each traps a portion of its parent's space-time fabric and thus inherits a balance of motion, mass, and dark energy.

Quantum systems, though small, behave like galactic systems because they trap fragments of the universal fabric directly. This is why they sometimes exhibit behaviors that appear random or non local: their closeness to the universal field gives them access to deeper laws. Yet, once they bond under gravity and form larger objects, they fall into the hierarchy of restrained systems.

This layered inheritance gives each system its distinct behavior. It explains why quantum particles vibrate near the speed of light, while planetary systems remain bound by gravitational fabrics.

5. Gravity and Black Holes

We discovered that gravity is not a force of attraction, but the space-time fabric's tendency to preserve its undisturbed three-dimensional geometry. Mass bends space-time, and space-time resists that deformation.

Where this resistance becomes extreme, a black hole forms. It is not a mystical singularity, but a region where the universal fabric reconnects with the ultimate fabric, dissolving trapped motion. Black holes anchor systems, stabilize fabrics, and may even exist in microscales within atoms or life.

Importantly, only primary and secondary systems, universal and galactic, are born with central black holes. Other systems, like stellar or planetary, have dense centers of mass but not true black holes, unless later formed by collapse.

6. Energy: Rest, Motion, and Transformation

The total kinetic energy of the universal system is in equilibrium with the resistance of the ultimate space-time fabric. Within this balance, all internal energies, mass, motion, and dark energy, must sum to cosmic harmony.

From this balance, we derived a powerful expression:

$$\frac{1}{2} mv^2 = mc^2 + DE$$

This equation told us:

- The universe itself has no direct mass because it has exhausted all dark energy to achieve ultimate motion.
- All mass and dark energy exist only in restrained child systems.

7. The Birth of Our Universe

When a galactic system escapes the grip of the universal system and nears ultimate velocity, it faces two choices:

dissolve into the stillness of the ultimate fabric or release its internal energy in a massive transformation.

When release occurs, the system does not erupt chaotically, but structurally. A new universe is born. In this way, universes give birth to universes, not from randomness, but from the pursuit of equilibrium. Each is like a particle emitted from a cosmic atom, carrying memory and momentum from its parent.

8. The New Universe Unfolds

Within a newborn universe, the first systems to appear are quantum systems, primary particles like photons and electrons. These are the seeds of everything else. Their motion, balance, and trapped energy define the new fabric.

From there, atoms, molecules, stars, and galaxies emerge. Each layer builds on the one before. Each system continues the same cycle of motion, resistance, and balance. At the center of every galactic system lies a black hole, a stabilizing defect in the fabric. These are not anomalies but necessities, evidence that motion has been restrained and restructured into orbit.

9. The Final Realization

Perhaps the most startling insight was this: life is not an exception. It also follows the same law. Where motion becomes complex and memory arises, life begins.

Motion is not only the story of galaxies and stars. It is the breath of living systems.

With this new lens, we can now look at familiar phenomena with fresh eyes. In the chapters ahead, we will apply the Universal Law to some of science's greatest mysteries:

Why do particles behave like both waves and points?

What truly happens in the double slit experiment?

Why do living systems grow, evolve, and die?

And where does consciousness come from?

We now carry a powerful torch. Let us take it forward into the unknown.

The decoding begins now.

Chapter 13: The Motion of Life

From this chapter onward, I am no longer presenting answers; I'm sharing possibilities. These are not conclusions backed by data, but thoughts that emerged when I followed motion to its deepest roots. They may be right, or they may not. But they arise naturally if we accept the logic of the Universal Law.

What I offer here is a continuation of that journey, a way of thinking, not a set of final truths. I am not a physicist. I'm not a biologist. I'm just someone who sat in silence and let the universe speak through motion.

If the Universal Law can explain time, space, gravity, and systems, then it must also have something to say about life.

So let's ask: What makes something alive?

We know by now that everything in our universe is a system created through motion. Galaxies revolve around black holes. Solar systems revolve around stars. Planets orbit suns. Electrons move around nuclei. The structure remains the same, only the scale changes. And at each level, motion is balanced by resistance: gravity, tension, space-time.

Life is also a system. But unlike a rock or a gas cloud, a living system does more than just persist.

It adapts.

It remembers.

It multiplies.

It responds.

It evolves.

Under the Universal Law, none of this is magical; it is simply a deeper level of organization and motion.

Every living organism is made of systems within systems. Cells are made of molecules, molecules made of atoms, and atoms made of subatomic particles. Each layer carries motion. Each layer carries structure. Life, then, is not the result of a mystical spark, but the natural consequence of complexity reaching a critical threshold.

And this complexity doesn't float in a void; it is shaped by the local structure of space-time. Everything we see on Earth, trees, animals, human beings, exists within Earth's space-time field. The way these organisms grow, behave, and interact is influenced directly by the fabric of this field, which is always trying to return to equilibrium. Life appears when that return is resisted, when something holds motion against the pull of reintegration.

Here is a new possibility: On planets, dark energy may often achieve equilibrium by converting into mass, the matter we recognize.

But what if some of this dark energy, instead of stabilizing into mass, continues to resist?

What if it transforms directly into dynamic, structured motion that cannot settle as static matter, what we experience as life?

This excess dark energy, which could not transform into ordinary matter, might manifest as the living pulse of a planetary system. It is not a ghost or anomaly; it is the same Universal Law, continuing its dance in a different form. In ancient traditions, this was perhaps called *chi*, or *prana*, the animating force that flows through all living things.

The first cell is always born in motion. The sperm moves. The egg waits. But the moment of contact is not just a collision; it is the transfer of trapped space-time into a new configuration. This new form resists the local gravitational push to collapse. And as the cell divides and grows, the energy we supply, sunlight, food, and water, gives it the ability to hold structure against this constant pressure.

At the heart of every life form lies a pulse, a rotation, a vibration. The beat of a heart is not just a function; it is a small, localized orbit. A system in stable motion. Much like a planet around a star, or a galaxy around a black hole. This sustained internal motion locks in a miniature version of the space-time structure, giving life its form and stability.

The cells that make up living beings are governed by the same Universal Law. They seek to reach their own ultimate velocity. But bound to the surface of the Earth, caught in its dense space-time field, they cannot escape. This proximity to a strong parent system imposes constraints, gravitational, energetic, and structural. It is this constraint that grounds life in more ways than one.

But here is where things change.

Life is not just orbital or inertial motion. It is purposeful motion. It seeks survival. It responds. It adapts.

Motion begins to remember.

Then it learns.

Then it adapts.

Then it predicts.

And finally, it reflects.

A single-celled organism swims toward food. A plant turns toward light. The brain reacts to danger. This adaptive motion arises from the structure, but it also begins to shape it. And that feedback loop creates entirely new behavior.

At a certain level of complexity, strange new things emerge.

Memory becomes possible: motion that can repeat itself.

Behavior becomes fluid: motion that changes with experience.

Intelligence appears: motion that can predict motion.

And then comes consciousness. Perhaps consciousness is not a thing, but a state. A moment when a system becomes so finely tuned to its own internal motion that it can perceive its own structure, not just react to the world, but reflect on itself.

For centuries, we treated life as an exception to the laws of physics. But through the Universal Law, life is no exception;

it is the inevitable outcome of structure, motion, and resistance reaching higher resonance.

And finally, there is death.

If life is the resistance of motion against collapse, then death is the collapse itself. The moment when the system no longer has the energy to maintain its structure. Motion breaks down. Order decays. And what remains dissolves back into the greater space-time field.

But even here, the Universal Law offers something deeper. Nothing is ever truly lost. Motion continues. Energy transforms. Structure returns.

Even consciousness, if it were truly bound to motion, may not end, but merely dissolve, like a ripple fading back into still water.

This, then, is how I see life through the lens of the Universal Law. Not as something outside of physics or beyond explanation, but as a natural consequence of motion, resistance, and structure reaching a higher order.

In the next chapter, we will dive into the world of time, a concept as elusive as life itself. But maybe it's not so strange after all. Maybe, at its root, it too follows the very same law we've been following since the beginning.

Let's go and find out.

Chapter 14: The Nature of Time

We often say time passes. That the future comes closer and the past slips away. But very few ever pause to ask,

What is time really?

Where does it come from?

Is it flowing, or are we?

If we look carefully, through the lens of the Universal Law, time reveals its true face. Time is not some invisible stream we are drifting through. It is not an external ruler measuring change. Time is not even a thing that exists independently. It is a consequence. A shadow. Time is the effect of motion, structured motion, moving through layers of space-time.

And just like all motion in the universe, time is layered, hierarchical, and directional.

Our universe, as we have already seen, is a system in motion. It has a velocity. And it has a direction. It is a vector. It cannot stop, and it cannot turn back. Everything inside it, galaxies, stars, planets, atoms, is riding this grand motion forward. And because time is born from this motion, time itself is also a vector. It, too, has a definite direction. It moves forward, and only forward.

Just as child systems inherit their motion from parent systems, they also inherit their experience of time. A planetary system moves more slowly than a galactic one, and thus time flows differently at that level. The deeper a system is embedded in the resistance of a parent space-time field, the more slowly its internal time moves. Time flows fastest in Universal Systems, where motion is pure and unresisted, and slowest in lower systems, like planets, where the structure is denser and more bound by gravitational fabric.

We experience time on Earth as a steady, even rhythm. It feels reliable, sunrise and sunset, aging and memory, ticking clocks. But this is just the rhythm of Earth's local space-time fabric, the speed at which we are able to resist and move through our parent system's structure. It's not the absolute pace of time; it's just our layer's experience of it.

We begin to notice the truth of this when we step outside our comfort zone. Take satellites, for example. When they move rapidly around the Earth, their velocity relative to us increases. According to Einstein's theory of relativity, their internal time begins to pass faster compared to ours. Their clocks run ahead. And indeed, our modern GPS systems need to constantly adjust those satellite clocks so they remain in sync with time on the Earth's surface.

From the Earth's frame of reference, satellite clocks appear to run faster due to their reduced gravitational resistance and increased velocity, but in their own frame, they tick normally. It is the difference in resistance through the parent field that causes this discrepancy. While traditional relativity considers both gravitational potential and velocity effects, within the framework of the Universal Law, we focus on resistance through the space-time fabric to explain the net flow of time.

But this isn't just some mathematical trick. It's a deeper truth: as the satellite increases its velocity, it starts approaching its ultimate velocity within Earth's gravitational system. The more it aligns with that pure motion, the less it resists the space-time fabric around it. And as resistance falls, time flows more freely. In other words, the faster you move relative to your parent system, the less space-time holds you back, and the faster time flows for you.

As a system moves closer to its ultimate velocity, it becomes more aligned with the parent field's motion, and its resistance drops. It flows more with time than against it; hence, time, as experienced internally, flows more freely.

Einstein was right, but now we see the reason behind the mathematics. It is not speed alone that dilates time; it is the relationship between a system's motion and the resistance of its space-time fabric. Time is a measure of how well a system is moving through its own layer of resistance.

Now comes the question people often ask when they hear about time being relative. If time can slow down, can it go backward? Can we reverse it, rewind it, build a machine that travels to the past?

The answer is simple and clear: no.

Because our universe is a vector, time is a vector too. It has one direction. That direction was set in motion the moment the Universal System was born. Motion has continued uninterrupted since then, and so has time. You can slow it down, you can stretch it out depending on your frame of reference, but you can never turn it around. You can never flip the arrow. You cannot go back.

Imagine trying to reverse your place in a river that is flowing from the mountains to the ocean. You could swim upstream, yes, but the river itself still flows downward. The water is not going to flow back up into the mountains. You are moving within the river, not controlling the river. The same is true of time. You can change your speed, but not your direction.

So all the science fiction stories, with their time machines and backward leaps through wormholes, with soldiers from the future returning to the past to stop wars or send killer robots, those are dreams built on misunderstanding. They may entertain, but they do not align with the reality we now understand.

Time travel to the future? Yes, in a way. If you move fast enough, you can experience less time than someone who stays behind. But you are not jumping forward, you are simply moving differently through the same river. When you return, more time has passed for others than for you. That's not magic. That's motion.

But can time travel to the past?

That breaks the vector. That asks the river to flow backward. That asks for motion to undo itself. And that cannot be. Because in the structure of this universe, motion never reverses, and time never flows backward.

And just as broken glass never reassembles on its own, the universe's motion cannot rewind. The arrow of time, once drawn, is irreversible, not only because of entropy, but because the structure of space-time itself forbids reversal.

What does all this mean for us, for life, for consciousness?

It means that we live in a forward-moving reality, where everything we are is built on the motion we've already taken. The past is not a place. It is a pattern, an imprint of motion now completed. It cannot be revisited. It can only be remembered. And the future? That is not a fixed destination waiting ahead. It is a path being built in real time, out of motion yet to happen.

Perhaps consciousness, too, is not a fixed trait, but a form of self-aware motion, one that moves forward in time just as our bodies do through space.

Time is not our enemy. It is not a curse. It is not an illusion. Time is simply the echo of the universe's movement. It is what we feel when structure resists motion, and what we experience as we move through this grand, unfolding sequence we call existence.

And just like everything else in this universe, time is governed by the Universal Law.

Reaching this understanding was never my job; it was simply my obsession. I am not a physicist. I am not a scientist. I am just someone who sat down one day and stared into space, into the stars, into silence, into motion, and refused to stop asking why. And somehow, just by following the motion, I found patterns. Patterns that began to make sense. Patterns that revealed a law.

But I have reached the limits of what a single, curious mind can do.

So from here, I want to hand the torch forward to those who have made it their life's work to investigate the unknown. To those who have the tools, the labs, the math, the methods.

In the coming chapters, I will attempt to express the Universal Law in more formal terms, mathematical, conceptual, and experimental. I will also offer interpretations of known phenomena through this lens, not as final answers, but as invitations for deeper inquiry.

This is no longer about me. This is about discovery. And if sitting in silence could lead me here, imagine what trained minds can build on top of it.

The universe is not hiding from us. It's always moving, always telling its story. **We just need to learn how to listen.**

Chapter 15: The Quantum Domain

From the very beginning, something kept tugging at my thoughts, an instinct, a quiet voice repeating the same question:

Does our size, our scale, change how we see the universe?

When we explored the early stages of the cosmos in Chapter 11, we began to understand how systems unfold through layers, each nested within the next. And as I reflected on this layered structure, I realized something simple yet profound: our human scale, our size relative to atoms, to electrons, to stars, may shape everything we observe. The way we perceive the universe is deeply tied to where we are in the hierarchy of systems.

To us, electrons appear strange. Their behavior is unpredictable. They seem to jump energy levels. They exist in clouds, not clear orbits.

But what if their behavior isn't strange at all, only strange to us?

Imagine a tiny organism, a living being the size of an electron, made of its own kind of matter, its own systems. For it, the electron wouldn't be mysterious. It would be normal. Just as we walk on Earth, perhaps it would revolve through electron-sized fields, governed by the same Universal Law, but at a different layer.

This idea always stayed with me: our observations are shaped by our position in the system. We are embedded within the space-time field of a planet, which is within a galaxy, which is within a universe, and our size, speed, and resistance to motion determine what we see and how.

Now, consider our Universal System again. Imagine it as a moving structure with a dark center, a black hole. Matter doesn't fall into it like a current. It gets left behind. The system keeps moving forward, and anything that can't keep up gets trapped in the curvature of the central void.

But viewed from a distance, from a cosmic, bird's eye view, this dynamic can be misleading. It can look like matter is moving in a straight line towards the center and disappearing. Around this central black hole, the space-time fabric curves and spirals, like the shape of a doughnut, or more precisely, like the circular fields around a conductor carrying current.

That image stopped me.

It's not that the matter is accelerating inward like an electric current, far from it. At the center lies a black hole, a remnant of a space-time rupture, where matter is no longer part of forward motion. But when you look at it from the outside, the illusion is there: matter vanishing inward, motion continuing outward, and a spiraling distortion around the center. A flow with resistance. A direction with tension.

It reminded me too much of **electromagnetism** to ignore.

In electromagnetism, we learn that when electrons move through a conductor, they create a circular magnetic field perpendicular to the direction of motion. Motion and field.

Line and curve. This geometry shows up again and again, from copper wires to planetary magnetic fields to galaxies. Could this be a coincidence? Or are we seeing the same behavior of the space-time fabric repeating across scales?

To me, it became clear: the electromagnetic effect may simply be a smaller-scale reflection of what the Universal Law predicts at cosmic levels.

What if electricity, magnetism, and even the strong and weak nuclear forces are not separate forces at all?

What if they're all expressions of one fundamental tendency: **the drive of space-time fabric to return to its undisturbed, natural form**?

We already understand that gravity is not a pulling force; it is resistance in the fabric when motion bends it. By the same logic, electric and magnetic forces could be curvatures of this same fabric, emerging in response to structured, confined motion. The tighter the bend, the stronger the restoring force.

Now think about **natural magnets**. These strange materials carry magnetic fields without any visible current. **Why?**

My guess is this: within natural magnets, certain particles may be moving at extremely high internal velocities. Their motion, while hidden from our instruments, might be bending the local space-time fabric of the planetary system more than ordinary materials can. This stronger curvature creates more resistance, and thus, the fabric pushes back harder. What we measure as a magnetic field is simply the space-time trying harder to restore its shape.

In that sense, the magnet is not generating a field; it's **fighting a deformation**. And the field is the side effect of that resistance.

This way of looking at it removes the artificial separation between electricity, magnetism, and nuclear forces. All of them become natural consequences of motion interacting with resistance. All of them become special cases of the same universal tendency, space-time trying to stay free, smooth, and undisturbed.

And now, this leads us to the quantum world, a world that has baffled physicists for over a century.

The quantum world is not a different place; it is our universe viewed at its smallest structural scale. It is the edge of our perception, where motion occurs with minimal resistance and structure becomes flexible, fluid, and ambiguous to our larger-scale instruments.

What if the uncertainty, duality, and strange entanglements of quantum mechanics are not so strange after all?

What if they are exactly what we should expect when motion occurs at the smallest possible scales, in systems with the least possible mass, and the least resistance?

If space-time bends more easily at smaller scales, then particles would not move in clearly defined paths. Their positions would blur. Their behavior would flicker. They would seem to be both here and there, because their motion is too subtle for large-scale systems like ours to lock down.

From our perspective, that's weird. From the Universal Law's perspective, it's just motion in minimal resistance.

I believe this quantum unpredictability is not a breakdown of logic; it is a breakdown of resolution. We are too large to observe the structure clearly. Like watching the fluttering of a leaf from across a valley, we see randomness where there may be a pattern, if only we were closer or smaller.

So maybe quantum mechanics isn't mysterious. Maybe it's simply **motion at the edge of perception**, governed by the same law that governs galaxies, stars, and life.

Even phenomena like quantum entanglement, where particles appear to influence each other instantaneously across vast distances, may not be a violation of causality, but a sign that their space-time fields were never truly separated in the first place.

But that's a deeper thought, and one I will explore more carefully in the next chapter.

For now, what matters is this:

Even at the smallest scale, the universe is still telling the same story.

Chapter 16: Entanglement

In the last chapter, we began peering into the quantum domain, a world that appears strange, elusive, and even chaotic from our human perspective. But we also realized that it may not be chaotic at all. Rather, it might simply be the same Universal Law acting at a much smaller scale, so small, in fact, that its behavior begins to blur before our eyes.

One phenomenon in this quantum world has always stood out, not just as strange, but almost mystical: **quantum entanglement**. When two particles become entangled, they seem to remain connected no matter how far apart they are. Change something in one, and the other responds instantly, even across vast distances. To classical minds, this breaks all rules. Einstein himself called it "spooky action at a distance."

But from the lens of the Universal Law, it isn't spooky. It's expected.

Let's begin not with particles, but with **systems**, because that's what everything is. Imagine two particles born from the same parent event. At their birth, they were not truly two things. They were one structure, one motion, within a shared field. Their space-time fabrics were never separate. They were part of the same undisturbed layer, together in motion, together in structure.

Now, imagine that this system stretches. The particles move apart in physical space. But the **space-time field that holds them doesn't tear**, it stretches with them. It still binds them through the invisible medium they emerged from.

To us, they look like two objects, two locations, two events. But underneath, they are **still part of the same ripple**, moving through a shared structure.

Think of it like pressing two coins into soft clay side by side, then pulling them apart. The surface now has two separate impressions, but if you look closely, there's a faint distortion between them, a stretched thread in the clay. They were pressed into the same medium. Touch one, and the other feels it, not because of magic, but because the medium itself is still connected.

That's what quantum entanglement is.

Not communication across distance, but **motion within an unbroken space-time association**.

This makes sense through the Universal Law. Systems that are born from the same motion remain linked unless something **resets** their structure. And in the quantum world, where resistance is low and motion flows with less disturbance, these connections can survive over great distances and timescales.

Entanglement, then, is not about particles "talking" to each other across space. It's about **shared tension within the same field**. When one side is disturbed, the other reflects that change, not because a message traveled, but because both are **still part of one unresolved motion**.

This is why, when one of the particles is measured or interacts with its environment, the entanglement often ends. The act of observation resets its fabric. It forms a new association, an updated system with new boundaries. The parent structure that once held both particles in unity is no longer preserved.

And this is also why entanglement is fragile at larger scales. Bigger systems have more mass, more internal complexity, and far more exposure to environmental resistance. Their original bonds dissolve quickly. But at the smallest scales, where motion is subtle and resistance is light, connections can persist longer, like echoes of an undisturbed birth.

This understanding isn't just philosophical; it's beginning to shape real technologies. In **quantum communication**, for instance, we use entangled particles to transmit secure information. Not because they send messages faster than light, but because their shared origin allows us to detect disturbances with perfect symmetry. The particles aren't transferring information; they're **remembering** the same motion.

All of this leads us to a simple truth:

Separation is an illusion.

We only see isolation because we observe from the outside, through the boundaries of larger systems. But under the Universal Law, systems are always part of deeper systems. And motion is always nested within layers of structure.

Entanglement is the echo of unity, of one system not yet fully divided, still holding together across the fabric of space-time.

If this is true, then we need to change how we think about interaction. Nothing is ever truly isolated. Every system carries the imprint of where it came from. And when we disturb one part of that system, the rest of it may still respond, even if we no longer see the thread connecting them.

This also means that what we call “information” may not need to be transmitted at all. It may already be there, embedded in structure, waiting to be revealed by motion.

Quantum entanglement doesn’t break the laws of the universe. It **reinforces them**. It is a quiet, elegant reminder that motion never forgets where it began, and space-time never forgets what it once held together.

In the next chapter, we’ll dive deeper into another great mystery: **wave, particle duality**. How can one object behave like both a wave and a particle? What is this duality really showing us about the layered resistance of the space-time field?

Once again, **the answers lie in motion.** And the deeper we look, the more clearly the law begins to speak.

Chapter 17: Wave, Particle Duality

As we descend deeper into the smallest scales of existence, the world begins to misbehave, or so we thought. Particles disappear and reappear. Light becomes both wave and point. Electrons seem to know we are watching. In the quantum world, certainty dissolves.

But through the lens of the Universal Law, these oddities begin to make sense, not as contradictions, but as consequences.

Let's begin with a simple, famous puzzle: the double-slit experiment.

When we fire a single photon or electron toward a plate with two narrow slits, we expect it to behave like a tiny marble. It should pass through one slit or the other, hitting a screen behind with a simple impact.

But that's not what happens. If we don't try to detect which slit it went through, the pattern it makes on the screen is not of two stripes; it is a series of bright and dark bands, like ripples of water. It's an interference pattern. Somehow, the single particle has behaved like a wave, interfering with itself as though it passed through **both slits** at once.

Yet the moment we place a detector to see which slit it went through, the pattern collapses. The interference vanishes. The particle acts like a particle again.

For decades, this has been explained as "wave-particle duality", a strange rule of the quantum world. But to me, this was never satisfying. It felt like a placeholder for something deeper.

So I sat and imagined, not the particle, but the **fabric it was moving through**.

Our Earth is not standing still. It is embedded in a larger system in motion, spinning, orbiting, flowing forward with the velocity of the Universal System. So everything fixed to Earth, the laboratory, the slits, the screen, is stationary in our local frame, but the **space-time fabric is flowing through it**, like wind through a canyon.

Now imagine this fabric, not empty and still, but constantly moving, bending, adjusting. A photon fired through the slits isn't creating its own wave; it is simply **moving through a medium** that already has flow and structure.

When that flowing space-time fabric encounters the two slits, it bends. It **distorts**, like water being forced through narrow openings. This isn't literal tearing, but a localized compression and redirection of flow. The two slit paths cause the flow lines of space-time to interfere with one another just after the slits, like eddies in a stream. These interference patterns already exist in the fabric **before** the photon arrives.

The photon simply follows this path. It is not "undecided" or "split." It is responding to the resistance and flow structure of the medium it moves through. Like any particle under the Universal Law, it seeks the path of **least resistance**, one that has already been shaped by the bending of the space-time field around the slits.

Now enter the detector.

The moment we place a measuring device near the slits, something large, physical, and gravitationally relevant, we warp the local space-time fabric. A detector is not just a passive tool. It has mass, structure, and its own trapped space-time field. Its presence introduces a **dominant resistance** that overwhelms the subtle flow pattern between the slits.

In doing so, the interference pattern in the fabric disappears. The flow lines no longer weave and interfere; they collapse. The photon, seeing no complex structure, simply takes a direct path through one slit or the other.

This explains why we cannot observe the particle and preserve the interference pattern: **the act of observation isn't magical, it is structural**. We are not collapsing a mystical wavefunction. We are collapsing the space-time flow field that the particle was responding to.

But then a question emerged in my mind: what if we placed the detector far away, **after** the photon had already passed through the slits and followed the flow path?

Could we still detect the photon's path without disturbing the pattern?

I believe we could.

If we placed a detector at a distance where it does not interfere with the space-time structure forming at the slits, the photon would still move through the original interference zone freely. Then, only afterward, we would collect information about its position. If done correctly, we might detect which slit the photon came through **without destroying the pattern**.

The photon is not conscious. It's not deciding based on knowledge. It is moving through structure, through a medium. What we call "wave-particle duality" is not a mystery; it is simply **motion responding to the resistance and structure of the space-time fabric**.

In regions of low resistance, motion spreads, behaving like a wave. In regions of high resistance, motion concentrates, behaving like a particle.

Just like a river narrows between rocks and then opens wide into a lake, motion expresses itself differently based on the resistance it encounters. Sometimes narrow, sometimes broad, but always in motion.

We once thought electrons were points. Then we imagined them as clouds. Then we treated them as probabilities. But perhaps they are none of these. Perhaps they are simply expressions of **structured motion**, shaped by the bending of space-time itself.

Wave-particle duality is not nature contradicting itself. It is nature being consistent. What changes is not the particle, it is the **field it moves through**.

These were the thoughts that convinced me there is motion, even when we cannot see it. That there is a medium, even if we cannot detect it. The evidence lies in the patterns. And those patterns repeat too clearly, too consistently, to be dismissed as coincidence.

The quantum world is not disconnected from reality; it is the same Universal Law, written at a smaller scale.

If invisible fabric flow can explain something as baffling as duality, what else might it explain? Perhaps the greatest mysteries of modern physics, those invisible forces that shape galaxies and stretch the cosmos itself, are not so mysterious after all.

Chapter 18: Misunderstood Universe

There are times when science, instead of admitting it doesn't understand something, invents something new to fill the gap. That is exactly what happened with **dark matter** and **dark energy**.

These terms weren't discovered; they were constructed.

Astronomers observed something odd: stars on the outer edges of galaxies were revolving far too fast. According to the equations we had from Newton and Einstein, this should be impossible. The visible mass in these galaxies wasn't enough to generate that much gravitational pull. To explain the mystery, scientists invented **dark matter**, an invisible substance, undetectable by any instrument, but supposedly making up most of the mass of galaxies.

Then came another shock: the universe isn't just expanding, it's **accelerating**. Galaxies are flying apart from each other faster and faster. So once again, to patch the math, scientists invented **dark energy**, a repulsive force that permeates space and pushes galaxies apart.

Between the two, dark matter and dark energy were said to make up **95% of the universe**, and yet we've never seen them, touched them, or measured them directly.

But perhaps we've misunderstood the data. Perhaps we don't need to invent invisible forces, because we've already uncovered the law that explains it all.

We've seen throughout this book that the **Universal Law** governs all motion in the cosmos. And that law is simple:

All systems tend to move in **uniform linear motion at their ultimate velocity**, the highest natural speed allowed by the structure of the universe.

We have already estimated the value of this **ultimate velocity** earlier in the book, derived from the motion of the Universal System itself.

Every system below that ultimate velocity, whether it's a planet, a star, a galaxy, or even a cell, will attempt to reach it, naturally. This effort to accelerate isn't driven by magic or mystery; it is simply the system trying to reduce the resistance it experiences from the space-time fabric and return to the undisturbed motion of Ultimate Space.

Let's look again at galaxies.

In their earliest stages, described in **Chapter 11**, galaxies formed as matter ejected from the Universal System's core. That matter was not stationary. It carried **inherited motion** from the Universal System, moving outward, with each fragment trying to stabilize itself as a new independent system.

So these galaxies **started their journey accelerating**. They weren't pushed apart by dark energy; they were already **born moving**, and the space-time fabric around them naturally allowed that motion to continue.

That alone would explain why galaxies appear to accelerate. There is no mysterious external push. There is only the natural unfolding of the Universal Law, **every object**

attempting to reach its ultimate velocity, converting mass into motion as gravitational resistance decreases with distance from its parent structure.

Now, what about the stars within galaxies?

Astronomers noticed that the outer stars, those farther away from the galactic center, revolve faster than expected. According to Newtonian gravity, they should be moving more slowly due to a weaker pull from the center. But they're not.

This, too, confirms the Universal Law.

Near the **center of galaxies** lies a **black hole**, the region where all motion **ceases**, where space-time fabric has fully collapsed, and resistance is absolute. As matter spirals inward toward this central point, it begins to **slow down**, losing its ability to continue in linear motion with the rest of the system. Eventually, it falls inward, not because it's being pulled, but because it has reached a standstill. It can no longer resist the overwhelming space-time pressure and is absorbed.

This is a unique case where **gravity doesn't accelerate** matter; it **decelerates** it. And that's a critical distinction.

In most systems, gravity accelerates objects **toward** a parent mass, like Earth pulling an apple. But when the parent mass is a **black hole**, it no longer accelerates. It **resists**. It slows. This is one of the few observable situations where we can **visibly test** the Universal Law's rule about motion and resistance.

Now contrast this with the stars **farther from the galactic center**. These outer stars are not bound by the intense resistance of the central black hole. Instead, they are **freer**, experiencing less resistance from the space-time fabric, and thus more capable of converting their mass into motion. That's why they revolve faster.

They are not being held in place by some invisible matter. They are simply **closer to achieving their own ultimate velocity**, fulfilling the Universal Law.

And the same idea explains the so-called dark energy.

As galaxies move farther apart, the gravitational resistance between them weakens. Each begins to behave more like an independent system, and thus, each continues accelerating toward its own motion equilibrium. What scientists interpreted as a strange, repulsive energy is simply **a system regaining its natural tendency**.

From this view, the galaxy isn't being pushed. It is **no longer being held back**.

So, what are dark matter and dark energy?

They are names we gave to our confusion.

They are attempts to describe a universe where matter and energy move, but our equations have failed to explain **why**.

But now, with the Universal Law in hand, **that all systems move toward uniform linear motion at their ultimate velocity**, the observations begin to make sense.

The stars are not breaking the rules. The galaxies are not defying gravity. They are all following the same principle:

As the **resistance of the space-time fabric drops**, motion **increases** because every system is trying to regain its place in the **undisturbed flow**.

This explanation not only accounts for the observed velocities but also offers a testable hypothesis:

In any system where the **central parent body is a black hole**, matter near the center should **move more slowly**, and matter farther out should **move faster**, with velocity increasing as gravitational resistance decreases. If we can consistently observe this across galaxies, it would be powerful evidence for the Universal Law.

We don't need to hide behind shadows to explain the cosmos.

We simply need to accept that everything is trying to move, freely, uniformly, eternally.

Dark matter and dark energy were placeholders. But now, the law is clear.

In the next chapter, we will go deeper into the question of **mass**, what it actually means to "have mass," how mass resists motion, and how it transforms into energy. Perhaps we will find that the **Higgs field** was pointing in the right direction, but the answer lies not in particles, but in **the structure of space-time itself**.

The journey of motion continues....

Chapter 19: Unexplained phenomenon

The strength of any law lies not only in its elegance but in its ability to explain the unexplained. The Universal Law, with its foundation in motion, disintegration, and reintegration of space-time fabric, offers fresh perspectives on many phenomena that remain either partially understood or wholly mysterious in modern physics. This chapter explores some of those phenomena and reinterprets them through the lens of the Universal Law.

Take the **Pioneer Anomaly**, where spacecraft appeared to slow unexpectedly as they moved out of the solar system. Early scientists were puzzled; they expected steady acceleration or uniform motion in deep space, not drag. But through the lens of the Universal Law, it becomes clearer. This spacecraft was not entering emptier space; they were transitioning from the Stellar/Solar system's disintegrated space-time field into the broader, higher-order galactic field.

And in this transition, their behavior would depend on which direction they exited. If they were moving away from the galactic center, toward the outer edges of the galaxy, they would align more with the galactic vector of motion. The reduced resistance and gravitational pull would allow them to accelerate more freely, moving closer to their own system's ultimate velocity.

But if the trajectory pointed toward the galactic center, where the central black hole exerts a strong space-time compression, then the system's resistance would increase.

The spacecraft, no longer supported by the solar system's tension and now moving toward the more intense curvature near the center, would slow down. Not because of a mysterious force, but because the fabric itself resists forward motion in that direction. It is, after all, where motion ceases. Where matter loses its ability to continue and collapses into stillness.

So the anomaly wasn't a mistake in measurement. It was a clue. A reminder that all motion is context, and the direction of transition between space-time hierarchies defines how motion behaves.

The Universe isn't empty. It's layered. And motion obeys the structure.

Take the **unexpected energy shifts** observed during **Earth flybys**, where spacecraft gained or lost speed in ways we couldn't fully explain. These variations can be seen as the result of the spacecraft temporarily aligning with or opposing the broader motion of solar or galactic systems. As it exits the gravitational well at different orientations, its movement through layered space-time fields causes subtle but measurable shifts in its velocity. Again, not magic, just motion realigning within a layered resistance structure.

Now, consider the **Casimir Effect.** The phenomenon shows that what we call a vacuum is not truly empty. In the smallest gaps between plates, pressure arises. Within the framework of the Universal Law, these fluctuations are not just virtual particles; they are the tension and interaction between trapped space-time fabrics attempting to rejoin their undisturbed parent field. A confined vacuum is simply a micro, resistance pocket of the Universal fabric, still expressing its drive toward unity.

The Cosmic Microwave Background, or CMB as they call it, is not some frozen echo from the Big Bang. That entire idea of a distant past light still wandering the universe doesn't sit right, not when we apply the Universal Law. What we're actually seeing is something much simpler and more active. The universe is massive, chaotic, and constantly in motion. Stars are exploding, systems are collapsing, and black holes are swallowing unimaginable amounts of matter. Every one of these events disturbs the space-time fabric. And by the Universal Law, any disturbance, any imbalance, must be corrected.

So how does the universe correct for lost mass?

It taps into its reserve: dark energy. This invisible field that surrounds everything is not just sitting idle. It plays a real role in maintaining the system. As mass disappears into black holes and instability builds, the space-time fabric responds by converting small amounts of dark energy into mass. That mass takes the form of low-energy photons, which we observe as the microwave background. This isn't leftover radiation. It's ongoing, continuous, and essential. These photons are being created all the time to maintain the balance of the system. From Earth, it looks stable because the visible universe is currently in a relatively stable phase of equilibrium.

The uniform microwave radiation we detect is nothing but this quiet correction mechanism. A constant background process where space-time, through dark energy, keeps replacing what's being lost. And here's where it gets interesting.

If one day, we observe that this microwave pattern has changed, even slightly, it means something has shifted. It could be a major mass event that we haven't yet detected, or something local like a tectonic stress building up. It means a new imbalance has formed, and space-time is responding.

This gives us a powerful idea. If we can measure local variations in this microwave radiation with enough sensitivity, we might be able to predict natural events before they happen. Earthquakes, black hole activity, or even sudden gravitational disturbances might all leave their signature in this background signal. The CMB isn't a leftover. It's a live signal. A diagnostic tool of the universe itself. The day it shifts is the day something bigger is at play. And if we learn how to read it properly, we'll be able to see the invisible forces shaping our world, before they make themselves known.

And finally, **one of the deepest mysteries: death.** We've discussed life as sustained motion, internal structure resisting collapse. So when that motion ceases, what remains? The trapped space-time fabric dissolves. Slowly, quietly, it begins reintegrating with the planetary field, then the solar, then the galactic. From structure to stillness. From individuality to unity. Death is not an end; it is a return. And if we truly wish to extend life, we must find ways to prolong or replenish the motion that sustains structure.

These are only a few of the questions modern science still wrestles with, and while this book does not aim to explain every anomaly ever recorded, the beauty lies in the simplicity of what it does uncover. With just one missing piece, the Universal Law, and the singular property of Ultimate Space-time Fabric, we've already unlocked a framework that makes sense of almost everything.

And perhaps, that’s all that was ever needed, **not to answer every question, but to illuminate the path forward with the revelation of a truth that was always there, waiting to be seen.**

Chapter 20: Future of Humanity

If the Universal Law truly holds, then our future is not just about survival; it is about transformation. For the first time in human history, we have in our hands not just a tool or a theory, but a revelation. A single understanding that links everything together: motion, matter, life, death, space, time, and even consciousness. And with that link comes a path to possibilities far beyond our current imagination.

Every problem humanity faces today, environmental destruction, energy shortages, disease, war, inequality, and even the fear of death, is born from our disconnection with the fabric we live in. We have spent centuries building technologies that defy nature, instead of understanding and aligning with it. But the moment we align with the Universal Law, everything changes.

Our energy problems, for example, are not problems at all. If all matter holds trapped space-time fabric, and all systems strive toward motion, then we are surrounded by potential. With this understanding, we can build energy systems that harness motion itself, gravity, inertia, spin, orbit, not to fight resistance, but to amplify it in constructive ways. We can unlock power from sources we once thought inert.

Our environmental challenges?

They come from trying to dominate rather than harmonize. Once we understand the Universal Law, we will stop building machines that burn and extract, and instead build systems that mimic nature's motion. Machines that orbit, flow, spiral, and systems that create no waste because they align with the motion and resistance patterns already present.

Even disease and aging can be re-understood. If life is sustained motion within a trapped space-time fabric, then preserving motion means preserving life. Medical science will shift from treating symptoms to preserving the structure and rhythm of internal motion. New tools will emerge, based on resonance, field manipulation, and tension balancing, that can heal not by chemical force, but by restoring flow.

Population?

Space?

Scarcity?

These are temporary illusions. When we understand the layers of motion and the ability to disintegrate and reintegrate fabric, we will learn to travel, first within our solar system, then beyond. The Universal Law does not restrict us to Earth. It opens the possibility to reach new worlds, not in centuries, but in decades. Motion will guide us to build vehicles that flow with the cosmic fabric, not against it. And when we learn to do that, borders will dissolve, not just on Earth, but across the stars.

Even war, the deepest wound of our civilization, will fade. When motion becomes our guiding law, when every life is seen as a unique structure of the universal field, when every person is a vessel of trapped space-time fabric striving toward the same goal, then division will seem pointless. We will no longer fight over land, for there will be planets enough. We will no longer fight over belief, for truth will be visible in motion.

And what of consciousness? It, too, is motion. It, too, is fabric. As we recognize this, we may find that our thoughts, when coherent, focused, and aligned, can influence the fields around us. Machines may one day respond not to touch or code, but to presence. We may build tools that sense our structure and respond to our motion. The interface between mind and matter will blur, not through fantasy, but through understanding.

The Universal Law removes the wall between what is observed and what is possible. It tells us that everything we see is born from a single principle, and that all systems, big or small, obey it. The future is not fiction waiting to be imagined. It is a field waiting to be shaped.

This book may have started as one man's curiosity, staring into the stars. But it ends as a call to action. I am not a scientist. I am not trained in laboratories or universities. But I have seen something true. And I urge the scientific community, physicists, doctors, engineers, thinkers, to read, to explore, to test, and to reach out. I am ready to explain everything I've seen. But more importantly, I am asking you to act. Act quickly. Act selflessly. Act for the betterment of humanity.

Because this is not just a discovery.

This is a turning point.

Let's end wars. Let's dissolve borders. Let's cure disease. Let's restore nature. Let's find new worlds.

Let's move, not just forward, but together, and finally, in the right direction.

Chapter 21: Mathematical Interpretation

After all the philosophical exploration, pattern recognition, and intuitive insights that have guided this book, it is now time to place the Universal Law into mathematical form. This chapter is not a final formulation but a beginning. These are the equations that naturally emerge from the principles we have uncovered so far. They are rooted not just in numbers, but in meaning, reflecting motion, resistance, and the drive of all systems toward their ultimate state.

This is the moment I never thought I would reach, the moment where the intangible becomes tangible. What I could not express in symbols for years, I now attempt with the help of something I never imagined would assist me, artificial intelligence. These mathematical forms are not mine in skill, but they are mine in spirit. They represent what I have felt all along but could never write alone.

We begin with the most foundational understanding: all systems strive to move in uniform linear motion at their ultimate velocity, unless resisted by the structure of the space-time fabric. This fundamental nature gives rise to energy, force, field behavior, and even life itself.

$$K_P(S) = M(S) + D(S)$$

- K_P(S): inherited kinetic energy of the system relative to parent fabric

- M(S) = mc^2: restrained portion expressed as mass energy

- D(S): dark energy (unfulfilled motion)

- V_u = sqrt(2)c: ultimate velocity at which D = 0

$$\frac{dM}{dt} = k_{res} R_f D - k_{free} F M$$

- M: mass energy

- R_f: restraint factor from parent fabric

- F: freedom factor

- k_res: constant for restraint converting D to M

- k_free: constant for freedom converting M to D

$$\frac{dD}{dt} = - k_{res} R_f D + k_{free} F M - k_{acc} F D$$

- D: dark energy

- k_acc: constant for acceleration converting D to kinetic energy

$$\frac{dK_P}{dt} = k_{acc}FD$$

- K_P: inherited kinetic energy
- FD: free portion of dark energy

$$\lim_{v \to 0} K_P = M + D \to 0$$

- At v → 0: motion completely restrained
- All M and D reintegrate into ultimate fabric

$$T_{local} \propto \frac{1}{R_f}$$

- T_local: flow of time for system
- R_f: resistance of parent space-time fabric

$$M_{life} > R_p$$

- M_life: sustained internal motion of living system
- R_p: passive resistance of environment

$$\Psi \approx \sum(m_i v_i e^{i\theta_i})$$

- Ψ: coherence of consciousness
- m_i: mass of constituent particle
- v_i: velocity of constituent particle
- θ_i: phase alignment of motion

$$K_P = M + D$$

- Fundamental relation at all scales
- Restraint: D → M
- Liberation: M → D → motion

Consider an example of a stone stationary on Earth's surface.

Stone of mass m = 10 kg: inherits Earth's parent motion at ultimate scale V_u = √2c ≈ 4.24 × 10^8 m/s.

Energy balance: K_P = 1/2 m V_u^2 = M + D
K_P = 8.99 × 10^17 J, M = 8.99 × 10^17 J, D ≈ 128 J. Almost all dark energy is restrained as mass.

Satellite of mass $m = 500$ kg: orbital velocity $v = 7{,}800$ m/s. $K_{Earth} = 1/2\, m v^2 = 1.52 \times 10^{10}$ J.

Relative to universal fabric: $K_{Universe} = 1/2\, m V_u^2 \approx 4.49 \times 10^{19}$ J.

Note: "Universe" means the ledger of the satellite relative to the universal fabric, not the mass of the universe itself.

These equations preserve energy across all scales by balancing motion, mass, and dark energy within parent fabrics.

They show why satellites orbit, why black holes dissolve systems, and why stones "at rest" still carry enormous kinetic energy.

They also hint at the possibility: by deliberately shifting the balance between mass and dark energy, we may design spacecraft that ride with fabric flow rather than fight against it.
Instead of brute force rockets, propulsion could one day be built on alignment, conversion efficiency, and resonance with the universal motion.

Chapter 22: Experiments

A scientific law becomes powerful only when it can be tested. The Universal Law described throughout this book is not meant to remain a philosophical insight; it is a framework to be measured, verified, and challenged. If true, it should reveal itself across scales, from quantum particles to galaxies.

This chapter outlines a suite of experiments and observational tests designed to probe the predictions of the Universal Law. Each is framed with a clear aim, method, prediction, and potential significance.

1. Field Tension Mapping Near Natural Magnets

Aim: Detect variations in space-time fabric tension around naturally occurring magnetic materials.

Method: Place natural permanent magnets (lodestone) and non-magnetic control rocks of identical mass in isolated vacuum chambers. Use interferometry, ultra-sensitive gravimeters, and atomic clock drift analysis to measure distortions in local field alignment. Rotate the chamber relative to Earth's galactic motion to test for anisotropy.

Prediction: Magnetized samples show measurable field tension differences not accounted for by conventional magnetostatics. According to the Universal Law, internal high-velocity particle motion bends the local fabric more strongly, producing a detectable restoring push.

Significance: Demonstrates that magnetic phenomena are not isolated forces but expressions of the same resistance field that governs gravity.

2. Orbital Time Flow Resistance Test

Aim: Test whether time dilation correlates with fabric resistance, not just velocity or gravity.

Method: Place two satellites at the same orbital altitude, one aligned with Earth's galactic motion and one opposed. Each carries an atomic clock compared against a reference on Earth.

Prediction: The aligned satellite moves with reduced resistance and its clock runs slightly faster, while the opposed satellite faces greater resistance and runs slower. Differences beyond relativity's predictions would validate that time is shaped by resistance through the fabric.

Significance: Expands relativity into a resistance-based law of time, matching the Universal Law.

3. High-Velocity Mass Conversion Test

Aim: Measure conversion of dark energy into effective mass at near-relativistic speeds.

Method: Accelerate neutral particles (silica beads, carbon clusters) in linear accelerators. Use interferometers and Casimir detectors along the tunnel to track field changes as the velocity rises.

Prediction: As particles approach their parent layer's ultimate velocity, field resistance decreases. Some apparent mass increase should actually resolve into dark energy conversion, altering the surrounding field tension.

Significance: Reframes relativistic mass increase as a fabric resistance effect, not just an inertial phenomenon.

4. Double-Slit Space-Time Flow Test

Aim: Validate that interference arises from structured fabric flow, not probabilistic collapse.

Method: Perform double-slit experiments with detectors both far downstream and directly at the slits.

Prediction: Distant detectors preserve interference. Near-slit detectors collapse the pattern, not from wavefunction collapse, but from local disturbance of the space-time flow field.

Significance: Redefines quantum uncertainty as structured motion in fabric rather than fundamental randomness.

5. Casimir Orientation Test

Aim: Detect orientation-dependent vacuum forces predicted by Universal Law.

Method: Place parallel plates in a vacuum chamber and measure Casimir forces with torsion balances at multiple orientations relative to Earth's galactic trajectory.

Prediction: Casimir force varies subtly with orientation. Plates against galactic flow show stronger vacuum pressure; aligned plates show weaker pressure.

Significance: Shows that vacuum fluctuations are micro-resistances in the space-time fabric, not random noise.

6. Black Hole Resistance Gradient Test

Aim: Validate that stellar velocities near black holes reflect resistance gradients.

Method: Use high-resolution telescopes (JWST, EHT) to track orbits of stars at different distances from galactic centers.

Prediction: Inner stars slow relative to Newtonian models, dissolving into stillness near the horizon. Outer stars accelerate beyond Newtonian predictions as resistance decreases outward.

Significance: Provides evidence against the need for dark matter while aligning galactic motion with the Universal Law.

7. Cosmic Microwave Background Drift Monitoring

Aim: Detect real-time changes in the CMB as signatures of equilibrium correction.

Method: Use ultra-sensitive microwave telescopes to track the CMB at fine temporal resolution. Look for subtle correlations with cosmic-scale events such as black hole mergers or galactic collisions.

Prediction: Small, localized variations should correlate with large events, showing the CMB is a live field adjustment, not just a relic of the Big Bang.

Significance: Redefines the CMB as the equilibrium "heartbeat" of the universal fabric.

8. Biological Resonance Alignment

Aim: Test if biological systems align with space-time resonance.

Method: Monitor cellular growth, neural oscillations, and heartbeat rhythms in varying resistance conditions (parabolic flights, ISS, centrifuge labs). Track coherence in quantum biological processes like photosynthesis and magnetoreception.

Prediction: Biological motion coherence shifts with changes in resistance, revealing life's sensitivity to deeper fabric layers.

Significance: Opens new frontiers in medicine and consciousness research rooted in motion and resonance.

9. Orbital Ledger Balance Experiment (New)

Aim: Directly test the energy balance equation KP=M+DK_P = M + DKP=M+D for orbiting bodies.

Method: Track satellites in low Earth orbit and compare their kinetic energy relative to Earth with their "ledger" relative to the universal velocity. Use ultra-precise gravimetry and energy accounting from maneuver corrections.

Prediction: Discrepancy between observed kinetic energy and universal ledger reveals the proportion of dark energy restrained as mass.

Significance: A real-world test of the mathematical framework from Chapter 21, bridging space technology with fundamental physics.

Final Perspective

These experiments are not just thought exercises. Most can be performed today with existing technology, from satellite clock comparisons to Casimir plates to black hole orbital surveys. If validated, the Universal Law will not only explain anomalies that forced us to invent "dark matter" and "dark energy," it will unify motion, mass, time, gravity, and even life under a single tested principle. The universe has given us patterns, and now it is time to test whether those patterns truly point to a single law.

Chapter 23: My Journey

This final chapter is not about the cosmos or equations. It is about me, the person behind these words. It is a glimpse into the human story that accompanied the birth of the Universal Law. A story not of a scholar or a professor, but of a dreamer burdened with a discovery too heavy to carry alone.

The journey started in my childhood, with a simple curiosity about the world around me. I often stared at the sky and wondered how everything was held together.

Why do things fall?

Why do stars shine?

But it was in 2010 that everything changed. That year, a thought struck me so deeply and clearly that it shook my understanding of existence. It wasn't an answer I read in a book or heard in a lecture. It was something I saw within. The missing link. The Universal Law.

What followed were not years of research in a lab, but a life lived in the real world. Between 2010 and now, I made several attempts to write this story, but life had its own demands. Responsibilities, survival, the need to put food on the table, and the daily chores of an ordinary man made it hard to find the time. Still, the idea never left me. In fact, with every scientific breakthrough announced over the years, my confidence grew. Each discovery echoed something I already felt deep within. I realized that the truth I had seen was not a fantasy; it was the framework behind what others were now beginning to glimpse.

In 2013, my journey took me to South Africa. I began working as a project engineer in the field of plant engineering. It was a long and difficult path. In 2017, I started my own company. I worked tirelessly, sometimes carrying material bags at construction sites due to labor shortages, other times returning home at 1 AM only to leave again at 6. There were moments of exhaustion, frustration, and doubt. But there were also friends, true companions who supported me through every storm.

Love entered my life, too. My first partner stood by me during the most difficult years, financially strained, mentally exhausted, and often emotionally unavailable. The stress of those times, coupled with my own shortcomings, made me drift apart and lose interest in nurturing what we had. Yet to this day, she remains a friend, offering support and understanding that I deeply respect and acknowledge. I dedicate this book, in part, to the troubles I caused her over the years. This is my subtle but sincere acknowledgment of her quiet strength during the most turbulent part of my journey.

My current partner entered my life when things had slightly stabilized. With fewer burdens weighing me down, I was able to nurture this relationship better and avoid the mistakes I had made in the past. She gives me the freedom to chase my wildest ideas and the strength to keep going.

I was never meant to live a conventional life. I'm a born entrepreneur, someone who created his own world, who never worked for anyone, who kept chasing a dream while the rest of the world settled.

Throughout my 11 years in South Africa, I kept trying to write this book. But another fear held me back. I knew that this discovery, if misused, could lead to destructive technologies. I didn't want that. I didn't want this knowledge to fall into the wrong hands. I wanted it to be a gift to humanity. That was always the dream.

And I missed my country, India. I always believed that my homeland was the right place to publish this discovery. That its spirit of knowledge and peace would protect and nurture what I had found. I felt it was my duty as a patriot to bring this gift home.

But how could I?

I was not a professor.

Not a scientist.

I had no stage or credentials.

Why would brilliant minds listen to me?

I was a nobody.

Still, the urge to share this truth was greater than my fear. And so, I began this book. With no publisher, no institution, no backing, only a mission. A mission to place this discovery in the open, to let the world see, verify, and build upon it. Even if only a small part of the Universal Law is true, it will reshape everything. Technology, physics, medicine, and our entire understanding of reality will leap forward. The dreams of Einstein, Hawking, and countless others will become achievable.

This may be the final chapter, but it is the beginning of something far greater than a book. It is the beginning of a revolution, not built on weapons or ideologies, but on a truth that belongs to us all.

I again emphasise: contact me, not because I hold all the answers, but because I hold a key.

Let us unlock the rest together.

Let us act before it's too late.

Let us stop fighting over land and resources and look up.

There are planets waiting.

There is space for everyone.

This was my journey. I offer it to you not as a story, but as a responsibility passed on. The future is not fiction; it is waiting to be shaped. And now, with this Law in hand, we are finally ready to begin.

Science had been confined for too long within the boundaries of orbital motion and the notion of trapped dark energy. But perhaps, with this gentle disturbance, the barriers have been lifted. Now, the dark energy within can be set free.

Now, the ultimate velocity of science, its true, unbounded motion, can finally be achieved.

Glossary

Universal Law

A foundational principle proposed in this book: *All matter or systems of matter are in uniform linear motion with ultimate velocity until acted upon by an external unbalanced force*. It is the unifying rule governing energy, gravity, time, matter, and consciousness.

Ultimate Space-time Fabric

The infinite, undisturbed, and structureless background through which the universe moves. It exists outside and beyond our observable cosmos and resists fragmentation or deformation. It is the source of the Universal Space-time Fabric.

Universal Space-time Fabric

A structured field created as a result of the universe's motion through the Ultimate Fabric. It binds all systems within the universe and is responsible for the emergence of forces like gravity and energy.

System

Any combination of matter and trapped space-time in motion. Systems are categorized by scale and hierarchy: quantum systems, man-made systems, planetary systems, solar systems, galactic systems, and the universal system.

Universal System

The entire observable universe, in uniform linear motion, is balanced within the Ultimate Fabric. It is the highest and most complete system known.

Galactic System

A galaxy, along with its central black hole and constituent systems (stars, solar systems), that has trapped a portion of the Universal Space-time Fabric.

Planetary System

A system composed of a planet and its natural satellites, created by the motion of a planet within a solar system.

Quantum System

The smallest known systems, like photons and fundamental particles, that trap portions of space-time and display high velocities close to the ultimate velocity.

Ultimate Velocity

The maximum velocity a system can achieve relative to the Universal Space-time Fabric, defined in this theory as $\sqrt{2} \times c \approx 424{,}264{,}068$ m/s. All systems strive toward this velocity to achieve perfect energy equilibrium.

Dark Energy (in this book's context)

A term used to represent the **latent potential** stored within systems that have not yet reached equilibrium. It is the *unconverted motion potential*, not to be confused with dark energy in standard cosmology.

Mass

Stored energy that arises when motion is restrained. Mass in this model is a *manifestation of unresolved dark energy*, i.e., energy that cannot currently convert into motion.

Gravity

A restorative push from the space-time fabric that acts to return the system to balance. Gravity is not attraction between masses but the result of space-time resisting deformation.

Black Hole

A region where the Universal Space-time Fabric reconnects with the Ultimate Fabric. Matter entering it loses its place in motion and dissolves back into the stillness of the ultimate field. It is not a point of infinite gravity, but a motion boundary.

Equilibrium

The state in which a system's kinetic energy matches the resisting energy of its parent space-time field. Systems naturally evolve toward this balanced condition.

Stored Energy

Energy that exists in the form of mass when motion is constrained. It can be released as kinetic energy when the system accelerates.

Hierarchy of Systems

The nested structure of motion systems in the universe each formed by trapping a part of the parent system's space-time fabric. Each lower system seeks equilibrium with its higher parent system.

Motion

The foundational cause of all energy and structure. In this framework, motion is not caused by energy; *it is the cause of energy*.

Resistance

The opposition faced by a system when it moves through a space-time field. It causes energy to be stored as mass or to emerge as gravity.

Time

A byproduct of motion through space-time. It is not absolute or external but emerges as a consequence of a system's resistance and velocity. Time flows in one direction, aligned with the motion of the universe.

Consciousness

Although not precisely defined in physical terms, it is proposed as *structured motion with self-awareness*. Consciousness may be related to internal black holes, like structures within living systems.

Reversal Threshold

A conceptual boundary where a system loses its motion relative to the universal flow, often associated with black holes or the death of systems.

Tribute

In Honor of the Thinkers Who Laid the Foundations

This book would not exist without the giants whose ideas, discoveries, and imaginations shaped our understanding of the universe. Whether I agreed with their conclusions or sought to go beyond them, their work created the very questions that led me to search deeper. It is with deep respect and gratitude that I acknowledge them here.

Sir Isaac Newton (1642–1727)

Major Works: *Philosophiæ Naturalis Principia Mathematica*
Father of classical mechanics. His laws of motion and gravity laid the groundwork for viewing the universe as a system governed by mathematical principles and external forces.

Albert Einstein (1879–1955)

Major Works: *Special Theory of Relativity* (1905), *General Theory of Relativity* (1915)
Redefined our understanding of space, time, mass, and gravity. His equation $E = mc^2$ is not only a pillar of physics but a stepping stone in this book's view of energy and motion.

Stephen Hawking (1942–2018)

Major Works: *A Brief History of Time*, *The Nature of Space and Time*
His work on black holes, singularities, and the origin of the universe brought deep questions to the mainstream and inspired many to reimagine the cosmos.

Niels Bohr (1885–1962)

Major Works: Foundational contributions to quantum theory and atomic structure.
His Copenhagen interpretation and model of the atom opened the door to a strange, layered universe at microscopic scales.

Werner Heisenberg (1901–1976)

Major Works: Originator of the Uncertainty Principle
Pushed science to acknowledge the limits of observation and precision, crucial for understanding systems at the quantum level.

Erwin Schrödinger (1887–1961)

Major Works: Schrödinger's Equation, *What Is Life?*
His ideas formed the bedrock of quantum mechanics and questioned the boundary between physics and biology, an idea central to this book's exploration of life and consciousness.

Max Planck (1858–1947)

Major Works: Originator of Quantum Theory
Planck's constant reshaped physics, showing that energy itself may come in quantized packets; motion, even at its base, has structure.

Galileo Galilei (1564–1642)

Major Works: *Dialogue Concerning the Two Chief World Systems*
Pioneer of modern observational astronomy and motion studies. His insistence on observation and questioning laid the path for empirical discovery.

James Clerk Maxwell (1831–1879)

Major Works: Maxwell's Equations
Unified electricity and magnetism into one coherent theory, demonstrating that even light is a product of motion in fields.

Richard Feynman (1918–1988)

Major Works: *Feynman Lectures on Physics*
A brilliant communicator of physics and one of the key figures in quantum electrodynamics. His playful curiosity lives on in any search for deeper laws.

Nikola Tesla (1856–1943)

Major Works: Contributions to alternating current (AC) power systems and electromagnetic theory
Visionary inventor who believed the universe was a system of energy, frequency, and vibration, resonating closely with this book's themes.

Carl Sagan (1934–1996)

Major Works: *Cosmos*, *Pale Blue Dot*
Popularized the idea that we are all part of a vast,

interconnected universe. His poetic awe for the cosmos inspires the soul of this work.

Subrahmanyan Chandrasekhar (1910–1995)

Major Works: Chandrasekhar Limit
His mathematical work on the life cycle of stars and the formation of black holes connects directly with this book's reinterpretation of cosmic balance.

Final Note

This book does not exist in isolation. It stands as a continuation of a long and often unacknowledged journey, one paved by countless individuals who dedicated their lives to understanding the universe.

Many of the names found in textbooks and biographies, those etched into history, have become synonymous with genius: they formulated the laws of motion, cracked open the atom, measured the stars, and questioned the very fabric of space and time. Their theories gave us tools to navigate the world, and their questions dared us to reach beyond what was known.

But beyond the celebrated few, there were, and still are, countless minds who worked in silence. Researchers who spent decades refining ideas, teachers who inspired questions, and inventors who failed a thousand times before others succeeded. There are thinkers whose papers were lost, whose voices were unheard, and whose names will never appear in any bibliography. Yet their quiet persistence formed the undercurrent of progress.

This book, and the **Universal Law** it presents, is built not just on breakthroughs but on struggles. Not only on brilliance, but on obsession, sacrifice, and vision. To all those known and unknown, whose contributions, dreams, and failures led us closer to the truth: **this is for you**.

While the Universal Law may reinterpret, challenge, or extend what has come before, It does so in humility, with full acknowledgment that **no discovery is born from nothing**.

Every page written here owes its existence to those who came before.

I do not claim to stand apart from them, only to stand **with them**, facing forward into the unknown. Without them, this journey could never have begun.

www.ingramcontent.com/pod-product-compliance
Ingram Content Group UK Ltd.
Pitfield, Milton Keynes, MK11 3LW, UK
UKHW060021051026
15591UKWH00006B/320
9798900895109